The CALC Handbook

Conceptual Activities for Learning the Calculus

Duane DeTemple

Jack Robertson

Washington State University

Dale Seymour Publications

Acknowledgments

The CALC Handbook was prepared in conjunction with the Conceptual Aids for Learning the Calculus Project, funded by the National Science Foundation Grant TPE 8751731. The Handbook materials were classroom tested by the following teachers associated with the project: Wyley Beatty, Lincoln High School, Tacoma, WA; Barbara Chamberlain, Shorecrest High School, Seattle, WA; Mel Griffith, Mead High School, Spokane, WA; Jim McLean, Kamiakin High School, Kennewich, WA; Shirley Ringo, Moscow High School, Moscow, ID. We wish to thank these teachers and their students for their contributions to the Handbook. We also thank Bekka Rauve, who conceived the illustrations, and the typesetters Marsha Bowmer, Kathy Hawbaker and Connie Pace. A special thank you to Connie Pace for the production of the camera-ready copy of this book.

Duane DeTemple and Jack Robertson
Washington State University
Pullman, WA 99164

Cover design: David Woods
Illustrations Concept: Bekka Rauve
Illustrations: Mitchell Rose
Technical art/mathematical figures: Scratchgravel Publishing Services

Copyright ©1991 by Dale Seymour Publications. All rights reserved. Printed in the United States of America. Published simultaneously in Canada.

This publication was prepared with the support of National Science Foundation Grant TPE 8751731. Any opinions, findings, conclusions, or recommendations expressed in this publication are those of the authors, and do not necessarily represent the views of the National Science Foundation.

Limited reproduction permission. The publisher grants permission to individual teachers who have purchased this book to reproduce pages as needed for use with their own students. Reproduction for an entire school or school district or for commercial use is prohibited.

ISBN 0-86651-543-7
Order number DS 21102

2 3 4 5 6 7 8 9 10 11-MA-95 94 93 92 91

Conceptual Activities for Learning the Calculus

THE CALC HANDBOOK

TABLE OF CONTENTS

Preface

The CALC Handbook–*Conceptual Activities for Learning the Calculus*–is meant to supplement any standard calculus text. The authors collected the material over many years, since teaching their first calculus course. The handbook was piloted in four high schools in the state of Washington for a year, then revised. Subsequently, approximately eighty high school teachers attended week-long training sessions in the content and pedagogy of the handbook. Explanations of topics accompany pages of exercises and solutions that you can photocopy for students. The "pick-and-choose" format allows total flexibility in your use of the handbook.

In this book, we have tried to include activities, carefully-chosen examples, and realistic applications that aid conceptual understanding. Wherever possible, we have emphasized a visual framework for the topic. There are numerous exercises that ask students to construct or interpret figures. Without such visual underpinnings, learning has to fall back on more abstract and easily-forgotten support systems.

Many activities emphasize the dynamic nature of the calculus. Most students' previous mathematical training has been finite, discrete, or static. Part of calculus's true beauty is that its analysis is often in motion. The definition of the derivative is a good example; another is the Intermediate Value Theorem and its applications. We have found that this dynamic view increases in importance as one continues to study mathematics, and is an enormous aid to understanding whatever is being studied. Both teachers and students should cultivate this perspective.

The examples, historical vignettes, and applications in this book are generally independent of one another, so you may present or omit what you choose. The sections vary considerably in scope, depth, and difficulty. In some cases they offer something "over and above" the usual for more advanced students. You can discuss most topics in one class period or less, or send them home as assignments after a standard presentation of the topic in class. A topically-arranged bibliography provides sources of further exercises or starting-points for the student projects.

Even though the sections are essentially independent, which allows for changes in order or deletions where appropriate, some sections develop a particular topic much more thoroughly than most calculus textbooks do. For example, a rich unit has been built around the harmonic series. By the book's end, the student has confronted many of the issues in series development as well as many topics that arise in numerical analysis, optimization, and probability.

For several reasons, we have placed minimal demands on the use of calculators and computers. The fact is, most of the essential concepts of the calculus can be taught well without this technology. Nevertheless, some of the applications, such as the materials on sequences and series, do assume that a calculator or computer is available. Also, a calculator or computer with graphing capability can augment the units on graphing, and you can adapt many other units to take advantage of a computer algebra system. All in all, the emergence of readily-accessible technology places increased emphasis on the student's conceptual knowledge and decreased emphasis on traditional paper-and-pencil computational skills. The *CALC Handbook* materials respond to these realigned emphases.

Two special features interspersed throughout the text help strengthen students' conceptual grasp of new topics: *The Big Picture* and *Probes.*

The Big Picture

As implied, these sections provide a broad overview of a topic. They are designed to introduce the topic in a general, non-technical way. You may choose to have students read these pages, or you may use them in introductory remarks about the topics. Each *Big Picture* is intended to tell students where they are headed and why–to provide the "lay of the land." It will often indicate how a topic fits into the larger fabric of mathematical study.

Probes

These sections consist of short questions whose answers can be given quickly, with minimal computation. The goal of the *Probes* is to refine, reinforce, and assess the student's conceptual knowledge. Many of the questions may be posed orally to the class.

GRAPHING

Sketching the Graphs of Composite Functions: One Step at a Time

To understand the nature and behavior of a function, it is very desirable to have a sketch of its graph. If the function is moderately complicated, even the powerful methods of calculus may seem of limited value, since the derivatives are likely to be as complicated as the function itself and information continues to be well-hidden.

Fortunately, the functions that often arise are composite functions, built up from such old friends as rational, trigonometric, exponential, and logarithmic functions. Since the graphs of the building-block functions are easy to sketch with qualitative accuracy, we can proceed step by step, one function at a time, to obtain an equally accurate graph of the more complicated composite function.

The method is probably explained best by taking an example.

$$f(x) = \frac{1}{[\ln(1 + \cos x)]^2 + 2}$$

$$\text{Thus } f(x) = f_6(f_5(f_4(f_3(f_2(f_1(x))))))$$

where f_1: cosine, f_2: add 1, f_3: logarithm

f_4: square, f_5: add 2, f_6: reciprocal

The "one step at a time" graph is now shown below, over one period interval $-\pi \leq x \leq \pi$

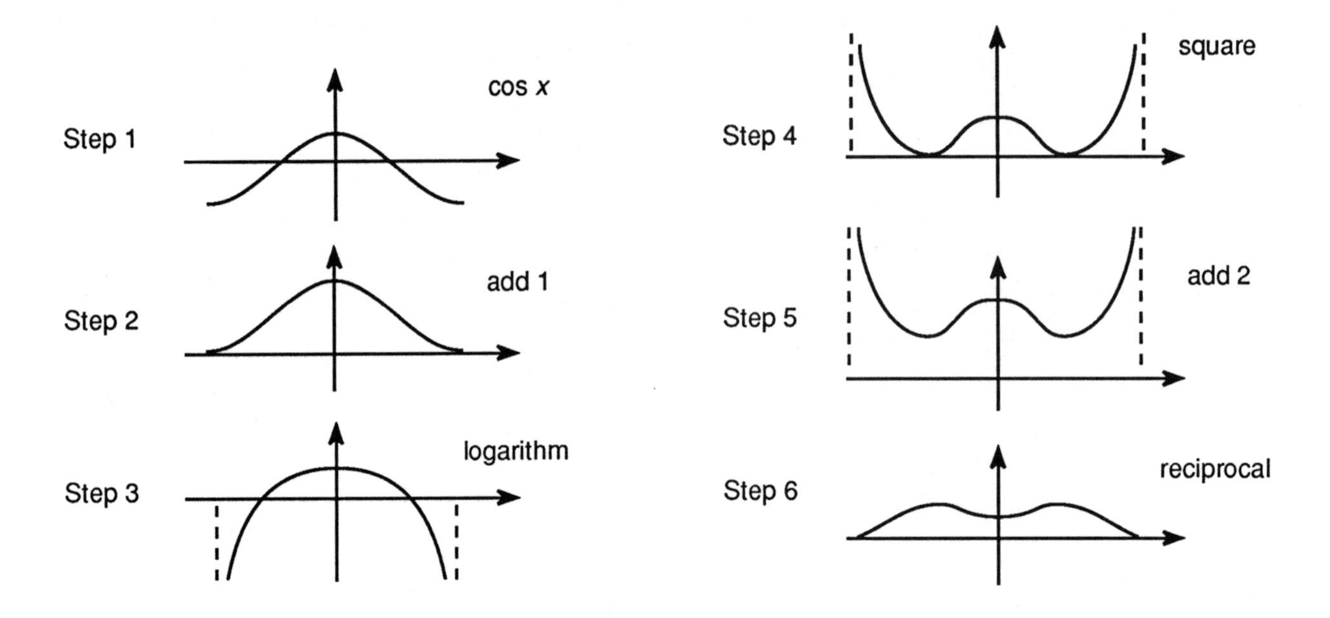

Exercises on Sketching the Graphs of Composite Functions

1. A sketch of a function f_1 is shown, and a second function f_2 is indicated below it. Graph $f_2(f_1(x))$.

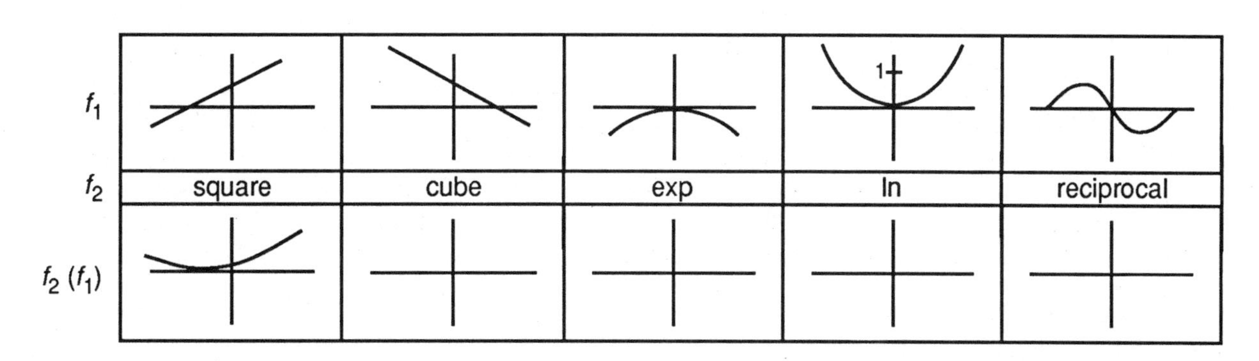

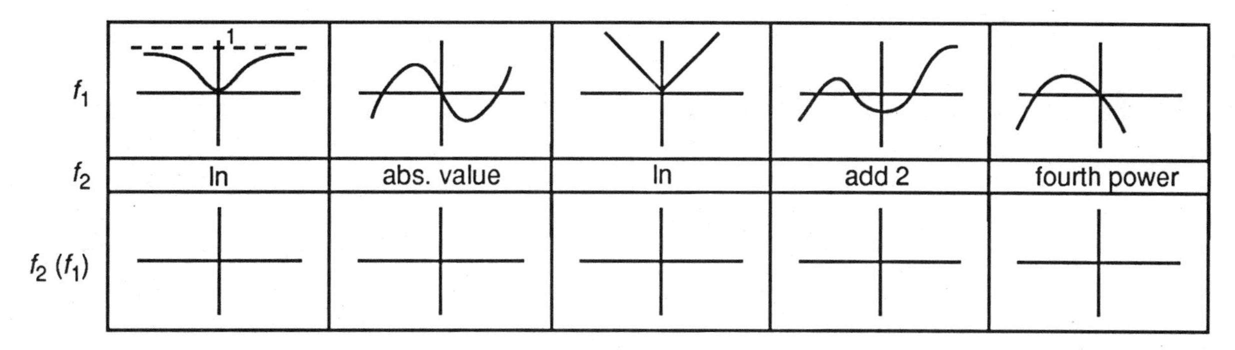

2. Use the "one step at a time" method to graph

$$g(x) = (|\, x^2 - 1\, | -1)^3$$

3. Graph

$$h(x) = 3e^{-\tan^2 x} \quad \text{for} \quad -\frac{\pi}{2} < x < \frac{\pi}{2}.$$

Reference: Keith Hirst, "Composite Functions and Graph Sketching," *Mathematics Gazette 72* (1988), 114-117.

Answers to Exercises on Sketching the Graphs of Composite Functions

1.

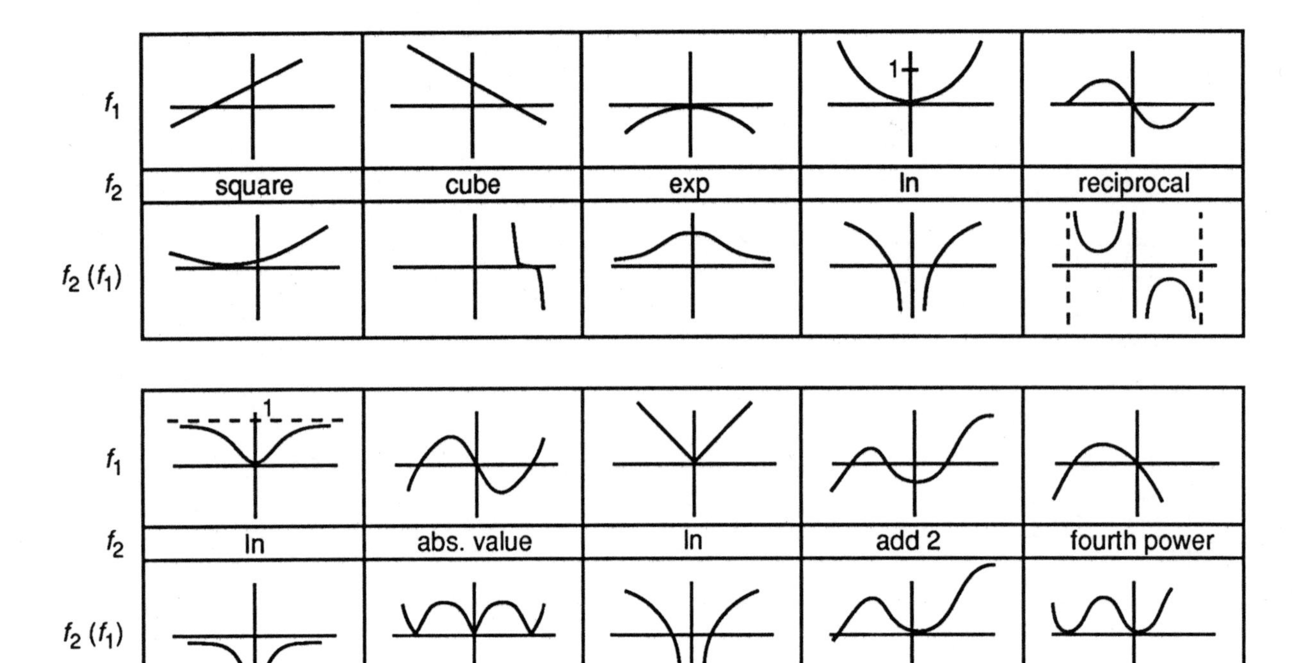

2.

Step 1: Square

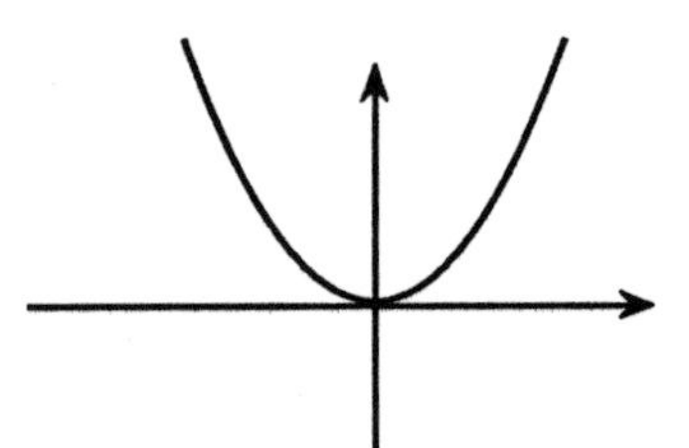

Step 3: Absolute value

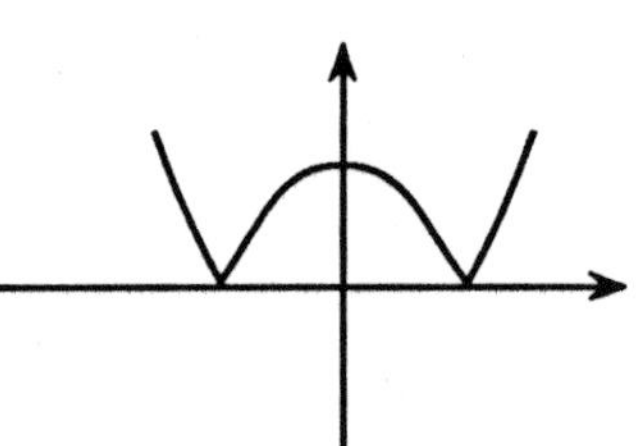

Step 5: Cube

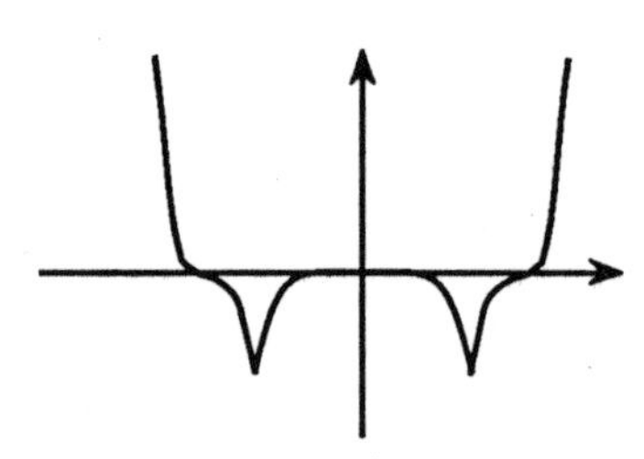

Step 2: Subtract 1

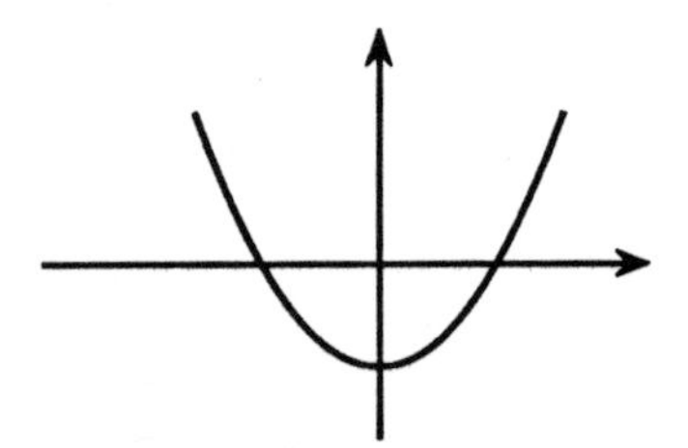

Step 4: Subtract 1

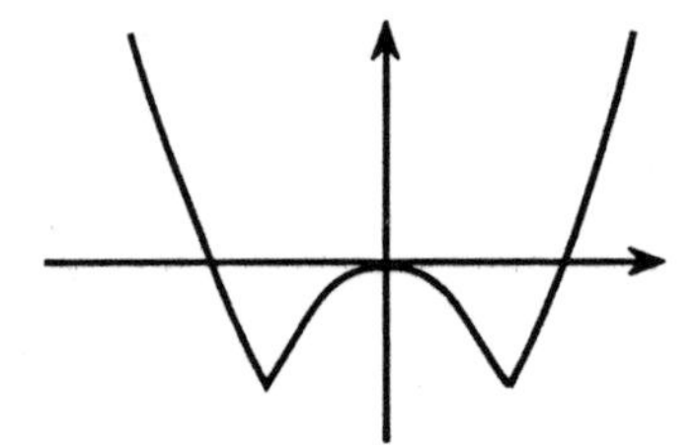

3.

Step 1: Tangent

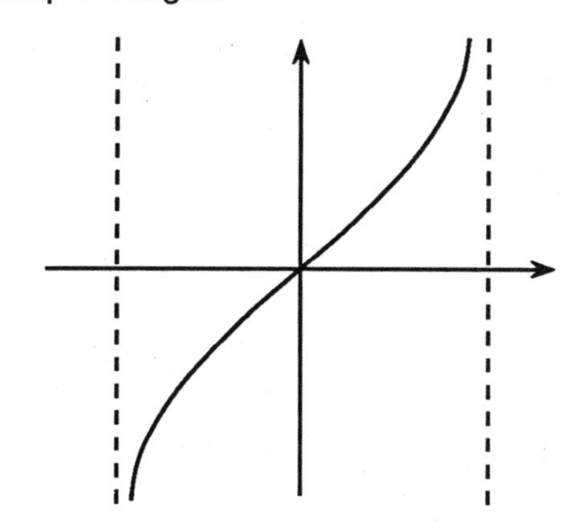

Step 2: Square

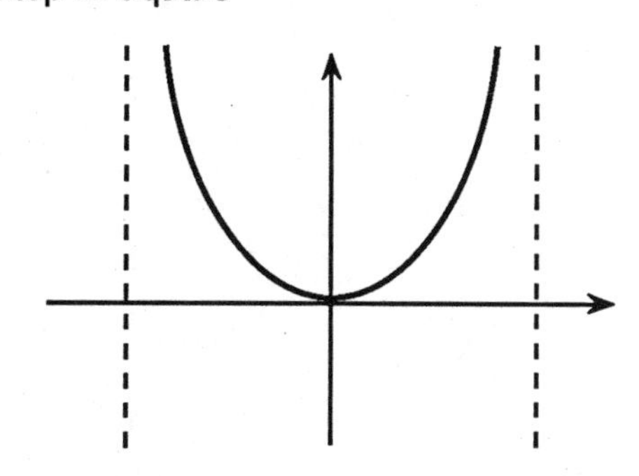

Step 3: Change sign

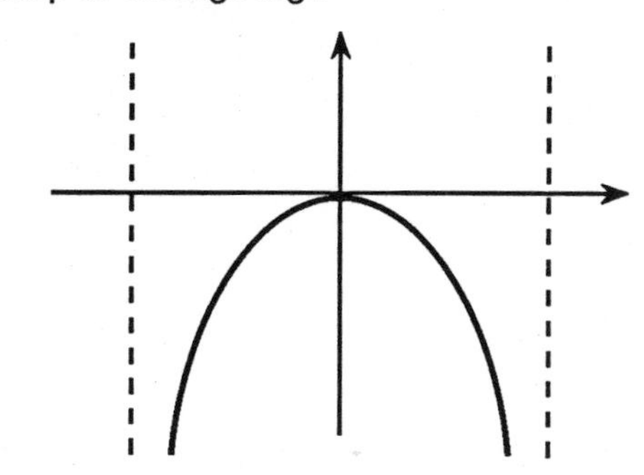

Step 4: Expontential

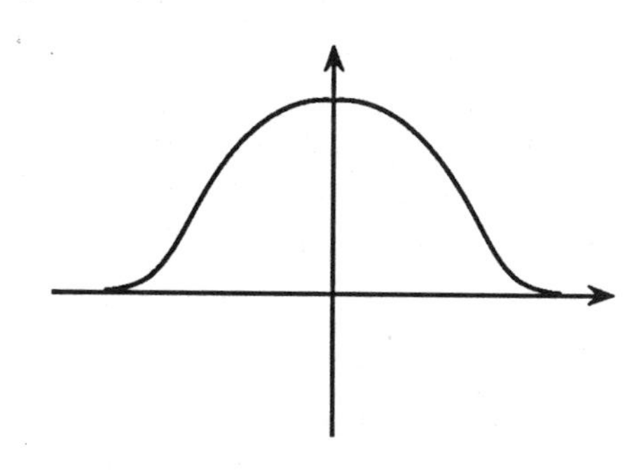

Step 5: Multiply by 3

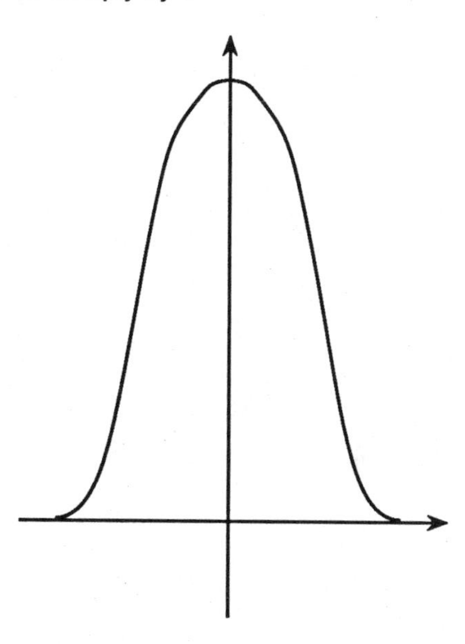

Plotting Curves in the Polar Plane

There is a very nice procedure for plotting curves in polar coordinates. One example will illustrate the method.

Plot $\rho = 1 + \sin 3\theta$ in polar coordinates.

θ	0	$\frac{\pi}{6}$	$\frac{\pi}{4}$	• • •
ρ	1	2	1.7	• • •

The usual method is to make a table like the one shown, plot those points, and hope to see what the curve should be.

Suppose we replace that table, which indicates just a small number of entries, by a "table" which shows *all* the values. We do this by graphing $\rho = 1 + \sin 3\theta$ in the familiar Cartesian coordinate system.

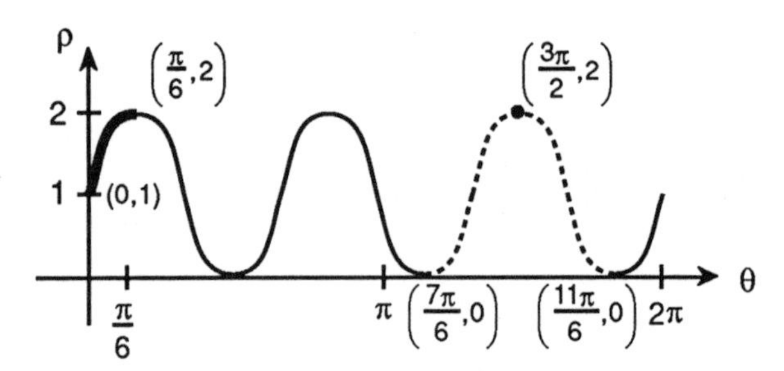

Cartesian graph of ρ = sin 3 θ

Finally, we use this graph just as we did the table above it; we read our (ρ, θ) values from the graph. The three entries we had in the table are labeled on the graph. But now we have more: we can see that as θ increases from 0 to $\pi/6$ then ρ increases from 1 to 2. This means the darkened part of the Cartesian graph (where $0 \leq \theta \leq \pi/6$) becomes the darkened portion of the polar graph shown between $\theta = 0$ and $\theta = \pi/6$.

Notice also that when $\theta = 7\pi/6$ and $11\pi/6$, ρ has value zero. Both $(0, \frac{7\pi}{6})$ and $(0, \frac{11\pi}{6})$ correspond to the pole in the polar plane. The dashed part of our Cartesian graph shows us that as θ increases from $\frac{7\pi}{6}$ to $\frac{11\pi}{6}$, ρ increases from 0 up to 2 (at $\theta = 9\pi/6 = 3\pi/2$), then decreases back to 0 again. As the angles swing from $7\pi/6$ to $11\pi/6$ in the polar plane, ρ will start at 0, increase to 2, and then decrease back to 0, which is shown by the dashed "petal" in the polar graph. Note that the portion of the Cartesian graph between each two successive points where the curve touches the horizontal axis corresponds to a "petal" in the polar graph.

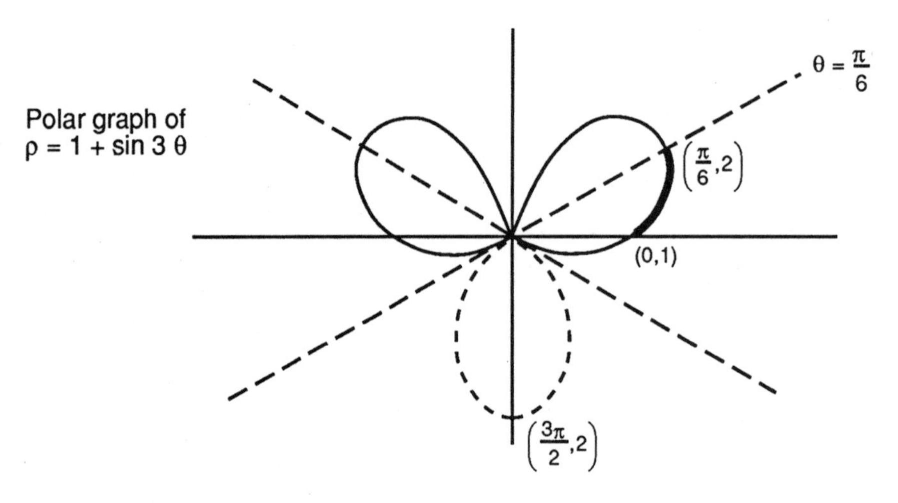

Polar graph of ρ = 1 + sin 3 θ

Here are two other examples:

1. Darken the part of the polar graph that corresponds to the dashed portion of the Cartesian graph.

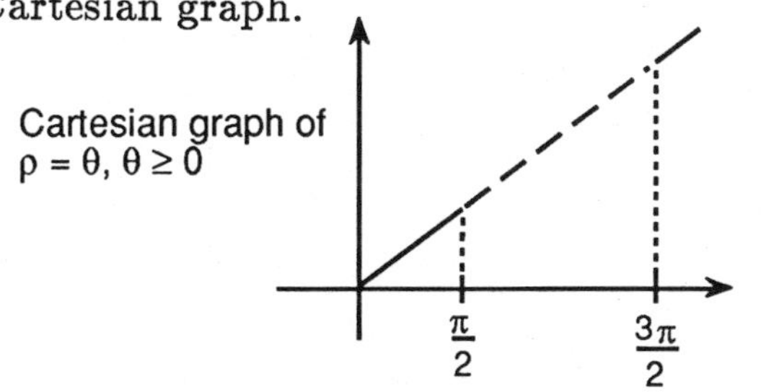

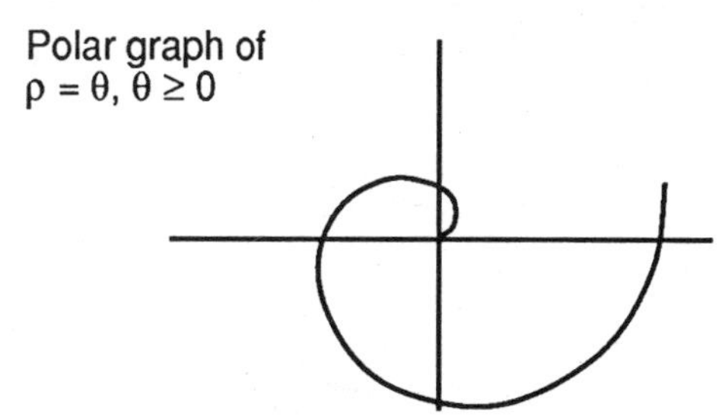

2. Darken the part of the polar graph that corresponds to the dashed portion of the Cartesian graph.

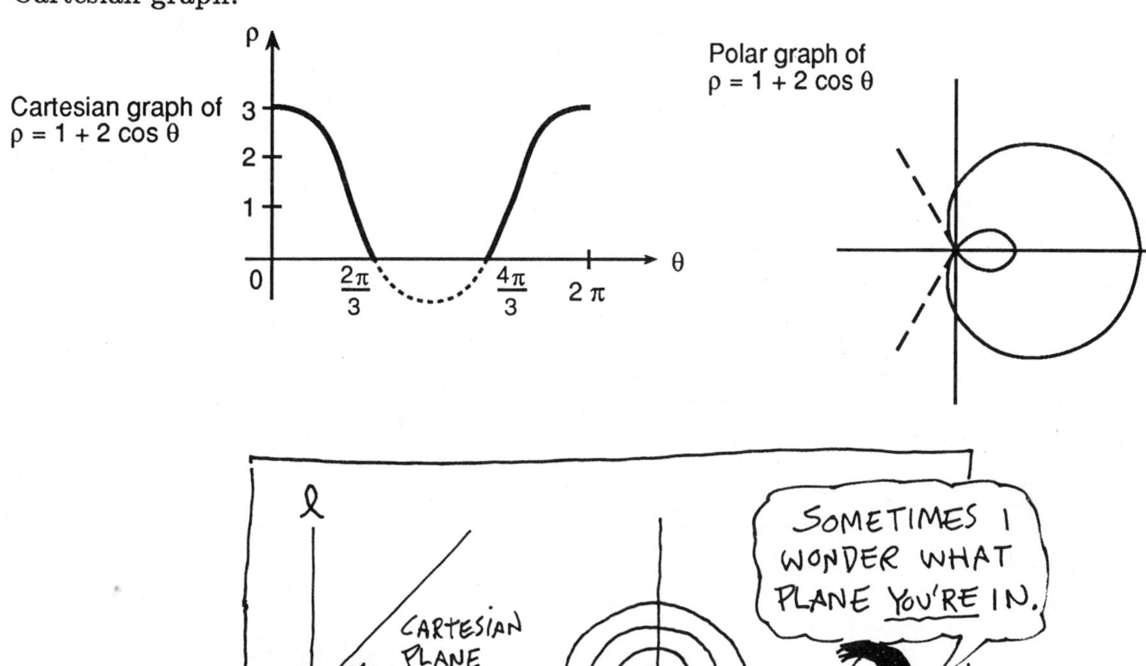

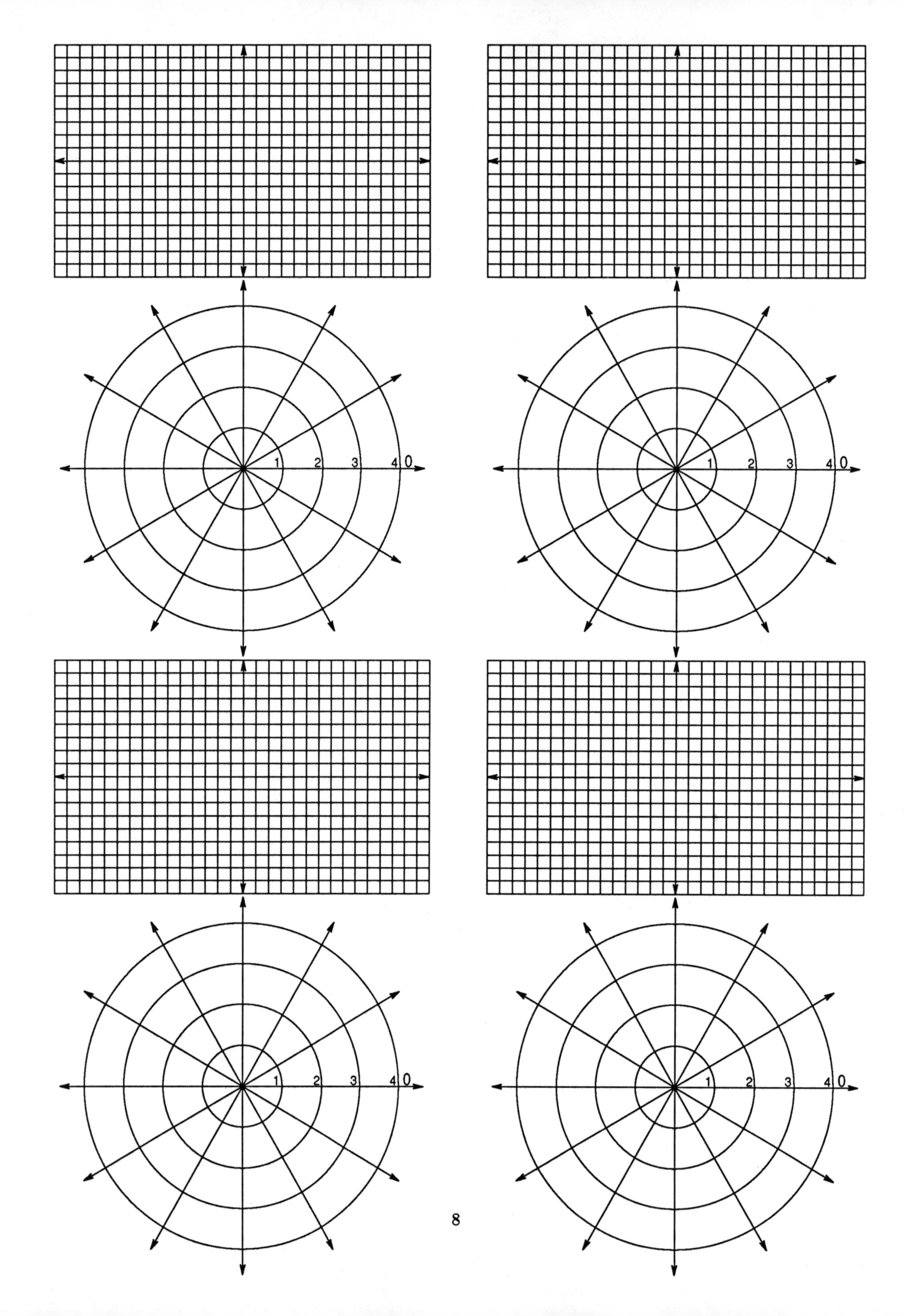
1
2
3
4
0

LIMITS

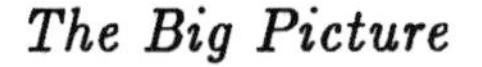

Limit–The Basic Element of Calculus

The first order of business in any treatment of calculus is to establish the notion of a limit. You will see the limit concept in three different contexts:

1. Limits will be used to define the *derivative*, the first tool of the calculus.
2. Limits of a somewhat different type than those used to get the derivative will define the *integral*, the second basic tool of the calculus.
3. *Infinite sequences and series*, which allow us to extend the process of addition to an *infinite* number of summands, require yet a third type of limit.

It will take time for all of this to unfold. These terms and processes will make sense to you later. In the meantime, *getting a good feeling for what a limit is should be a high priority for you at this early stage of your study of the calculus; it won't go away.* You should also understand that the limit concept is not simply an abstract concept of interest only to the theoretically-minded. In fact, there are many contexts (for example, engineering, environmental science, population studies) in which the limit concept plays a decisive role in analyzing and solving a practical, real-world problem.

One interesting thing to realize is that we encountered infinite series in about the 4th grade when we wrote $\frac{1}{3} = .333\cdots$ What in the world does that mean, $\frac{1}{3} = \frac{3}{10} + \frac{3}{100} + \frac{3}{1000} + \cdots$? We never added an infinite number of terms before. Most of us don't recall being bothered by it in grade school, where it seemed innocent and natural enough, but when we reach the calculus we find what a huge (but gentle) sleeping dog this is. Most people think $\frac{1}{3} = .333\cdots$ is okay, but they balk at $1 = .999\cdots$. You can't have one without the other. Making rigorous sense of all of this is something you can look forward to when you learn about infinite series.

A little friendly encouragement is in order. For a while the definition of the limit will likely seem awkward, maybe even mysterious. What is all this epsilon and delta stuff, anyway? Don't despair–you are in good company, because most people have the same problem. We hope the time will come soon, after some experience using the definition, when you will see the beauty of the statement's precision. And after you see how the limit is used to get the derivative and integral, you will realize that your mathematics has taken on a brand new dynamic and "continuous" character that is different from your previous mathematical tools. These new, more sophisticated tools are exactly what gives the calculus its extraordinary power to solve problems that otherwise would seem hopelessly difficult to solve.

Visualizing Epsilon and Delta

Central to using limits are ϵ, δ arguments. The game is played as follows. We are challenged with an arbitrary positive number ϵ and must then produce a corresponding positive number δ that does "appropriate things." We usually think of ϵ as small (if we can find δs for small ϵs, then we can find them for larger ones, as we will see below). Paul Erdös, a famous mathematician, refers to all small children as "epsilons." Review the following:

> *Definition*: $\lim_{x \to a} f(x) = L$ means that for every $\epsilon > 0$, a $\delta > 0$ can be found so that for any $x \neq a$ in the interval $(a - \delta, a + \delta)$ we have $L - \epsilon < f(x) < L + \epsilon$. (Study the figure. Remember the ϵ comes first.)

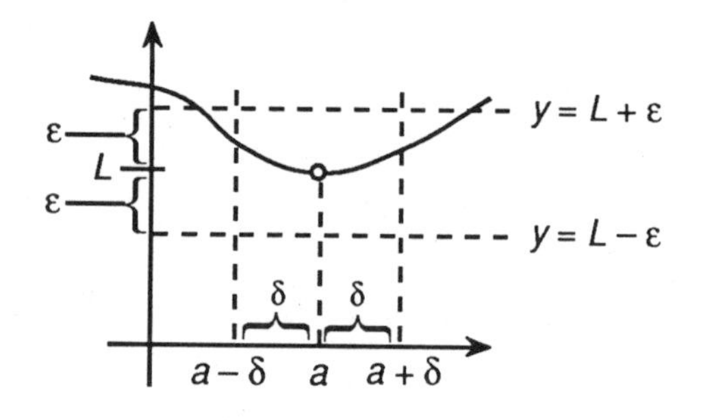

The ϵ, δ process can be visualized as a game of lines. If you are given two horizontal lines $y = L + \epsilon$ and $y = L - \epsilon$, you must then produce two vertical lines $x = a - \delta$ and $x = a + \delta$, so that the rectangle formed by the four lines has the property that the graph stays in the rectangle for all x's between $a - \delta$ and $a + \delta$, except possibly at $x = a$ itself. You get to choose $\delta > 0$.

Epsilon and Delta

1. Assume that f and a are given, and L is the limit of $f(x)$ as x approaches a.
   a. If for $\epsilon = \frac{1}{4}$ we are able to choose $\delta = \frac{1}{2}$, what other values of δ could be chosen?
   b. If for $\epsilon = \frac{1}{4}$ we are able to choose $\delta = \frac{1}{2}$, for which other ϵ would $\delta = \frac{1}{2}$ work?

2. Assume that f and a are given. In general, as ϵ gets smaller, δ gets (larger/smaller).

3. Assume ϵ is given and for that ϵ we want to find corresponding δ at two points a and b. In general terms, if the graph of f is steeper at a than it is at b, then the δ at $x = a$ is (larger/smaller) than the δ at $x = b$.

4. Give an example of a function such that for any point a and any $\epsilon > 0$ the corresponding δ must always be smaller than ϵ.

5. Give an example of a function such that for any point a and any $\epsilon > 0$ the corresponding δ may always be chosen larger than ϵ.

6. Suppose for *any* point $x = a$ and *any* $\epsilon > 0$ I can always choose $\delta = 1$. What can you say about f?

Answers to Epsilon and Delta Probes

1. (a) Any δ with $0 < \delta < \frac{1}{2}$; that is, any smaller positive δ will work.
   (b) Any ϵ with $\epsilon > \frac{1}{4}$.

2. Smaller.

3. Smaller.

4. $f(x) = 2x$ is one example.

5. $f(x) = \frac{1}{2}x$; δ could always be 2ϵ.

6. f would have to be a constant function. If f is not constant, then there is an x_1 and x_2 with $| x_1 - x_2 | < 1$ and $| f(x_1) - f(x_2) | = a > 0$. Choose $a = x_1$ and $\epsilon = \frac{a}{2}$. Then δ can't be 1.

The Big Picture

Continuous Functions

When you carefully examine the definition for the function f to be continuous at $x = a$, you will see that it is closely related to the limit of f existing at $x = a$. The only difference is that the test for the limit does not examine the value of f at $x = a$, and in fact f may not even be defined at $x = a$, and yet the limit exists. To be continuous at $x = a$, f must first have a limit and then meet two additional requirements. Let's be specific.

The function f is continuous at $x = a$ provided:

1. $\lim_{x \to a} f(x)$ exists,

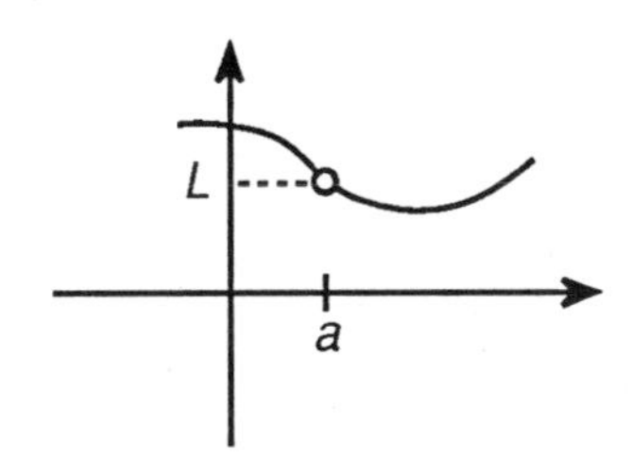

Think, roughly, that f behaves well near a in the sense that $f(x)$ is near L when x is near a. Note that there is no requirement for what f does at $x = a$.

2. f is defined at $x = a$; that is, $f(a)$ makes sense because a is in the domain of f, and
3. $\lim_{x \to a} f(x) = f(a)$.

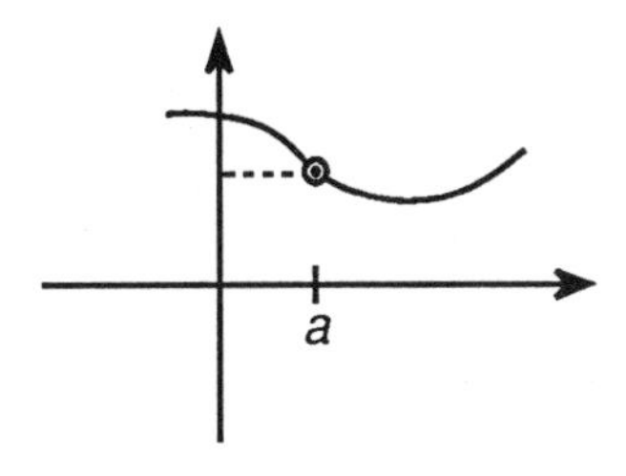

f takes on the value at $x = a$ that fills the blank circle we had in condition 1. Condition 3 requires the two numbers guaranteed by conditions 1 and 2 are equal.

Operations with continuous functions are at the heart of the calculus. We will be finding their maximum and minimum values (using derivatives), we will be finding the areas under their graphs (using integrals), and all of this will provide us with an impressive variety of applications. Continuity is an important concept because if the function has discontinuities, then one must hope that alternate methods can be found that take the discontinuities into account. Many times when a function fails to satisfy a continuity assumption, the result is false.

Notice that the definition of continuity is a point-by-point property of the function. There are functions that are defined everywhere but only continuous at a single point of the domain; there are other functions that are continuous at all points except one (and in some of these cases the function would become continuous there, also, if it were only redefined at that one point); there are even functions whose domain is the whole real line that are continuous at the irrational numbers and discontinuous at the rationals (but, interestingly enough, not vice-versa). Luckily, the more complicated situations rarely occur in any real applications and so they won't be a worry to us now.

When you think of continuous functions, you should get in the habit of acting like a botanist and classifying them whenever possible. Some of the most important "genres," all of which you have worked with extensively already, are:

1. *Polynomials*

   These are functions of the form $f(x) = a_0 + a_1x + a_2x^2 + \cdots + a_nx^n, a_i$ real numbers, which are defined and continuous everywhere. Note that constant functions $f(x) = c$ and linear functions $f(x) = ax + b$ are polynomials. Polynomials are the simplest of all continuous functions; their value at any point of the domain can be determined by a finite number of multiplications and additions. Thus, calculators and computers love polynomials. In fact, when you ask a computer for the value of $\cos 2.167$ or $\log 6.29$, it usually evaluates a *polynomial* that is a good approximation of the cosine or log function near 2.167 or 6.29, respectively, and calculates the value of that polynomial at the point.

2. *Rational Functions*

These are functions of the form $f(x) = \frac{p(x)}{q(x)}$, where p and q are both polynomials. They are defined and continuous wherever q is not zero.

3. *Trigonometric Functions*

The two most important families of trigonometric functions are those of the form $f(x) = a\sin(bx + c)$ or $g(x) = a\cos(bx + c)$, which you should know how to graph. Thus you know their amplitudes, periods, and where the graphs cross the x-axis or take on maximum or minimum values. The tangent family $f(x) = a\tan(bx + c)$ is a quotient of sine and cosine, and you should also be able to graph a member of this family. The remaining three trigonometric families are reciprocals of the three above. Any study of periodic behavior–for example, vibrating springs, vibrations in buildings or bridges, or any repeating seasonal phenomena (just to name a few)–requires the use of trigonometric functions. Sine and cosine are continuous everywhere; you should be able to locate the discontinuities of the others.

4. *Logarithmic and Exponential Functions*

You have studied the class of exponential functions $y = Ab^{kx}$ and the family of their inverses, the logarithmic functions $y = C\log_b kx$. You should know the general characteristics of these functions, the shape of their graphs, and how different values of the constants A, b, k, etc. influence the nature of these functions. The number b is of special importance: it must satisfy $b > 0$ and $b \neq 1$, and is known as the base. The value of b which appears repeatedly is the irrational number $e = 2.718\cdots$. Applications of this class of continuous functions include population studies, investment behavior, and radioactive decay.

One very important and very nice property of continuous functions that we use all the time is that the sums, differences, products, and quotients of continuous functions are again continuous (except where the denominator in division is zero). Also, the compositions of continuous functions are continuous.

Continuity

Recall the three conditions that make a function continuous at $x = a$.

1. $\lim_{x \to a} f(x)$ exists (this tells us something about the behavior of f for values of x near a.)
2. f is defined at a (that is, $f(a)$ makes sense; we know f has a value at $x = a$.)
3. $f(a) = \lim_{x \to a} f(x)$ (the two numbers insured in 1. and 2. are equal.)

1. Let $a = 0$.

   (a) Draw the graph of a function where condition 1 is met but 2 isn't.

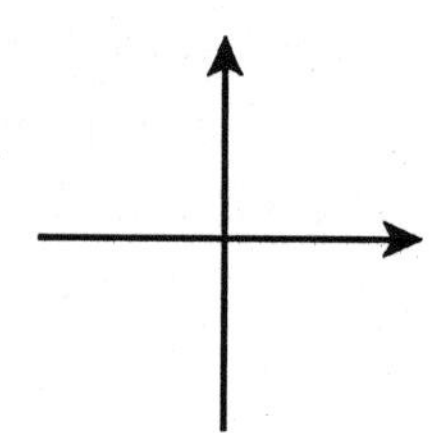

   (b) Draw the graph of a function where condition 2 is met but 1 isn't.

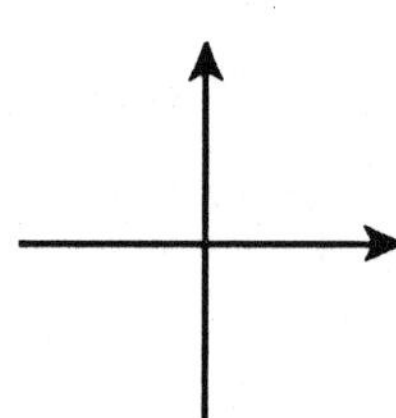

   (c) Draw the graph of a function where conditions 1 and 2 are met but 3 isn't.

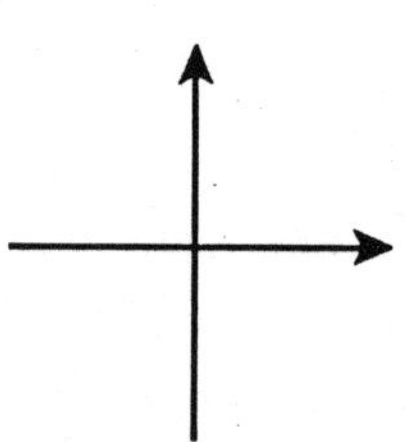

2. Draw the graph of a function that is continuous except at a single point, but would become continuous everywhere if it were redefined at that point.

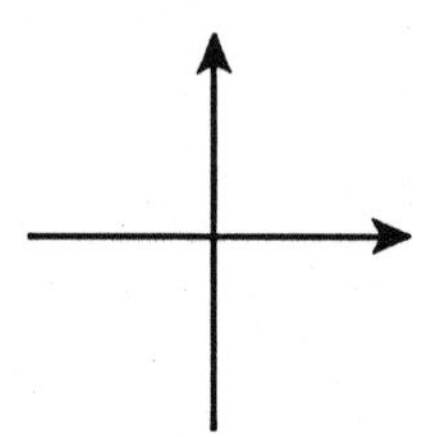

3. Draw the graph of a function that is continuous except at a single point, and could not be made continuous there regardless how the function was redefined at that point.

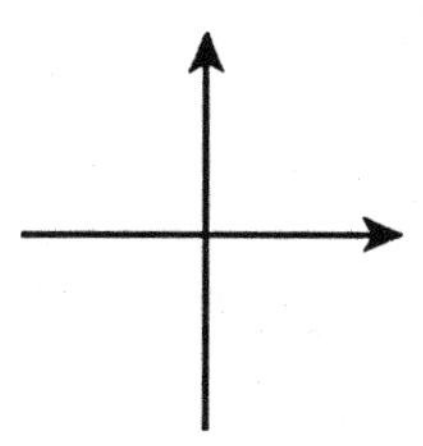

4. (a) If f is continuous everywhere and $f(0) = 0$, can $f(0.00001)$ be as large as $1,000,000$?
   Explain.

   (b) If f is continuous everywhere and $f(0) = 0$, must there be some $\delta > 0$ so that $-10^{-6} < f(x) < 10^{-6}$ for $-\delta < x < \delta$?
   Explain.

5. If $-1 < f(x) < -.5$ for $0 < x < 1$ and that is all we know about f,

   (a) could f be continuous at $x = 1$ if $f(1) = 0$?
   (b) could f be continuous at $x = 1$ if $f(0) = 0$?
   (c) could f be continuous at $x = 1$ if $f(1) = -1$?
   (d) could f be continuous at $x = 0$ if $f(0) = -0.75$?
   (e) could f be continuous at $x = .5$ if $f(.5) = -2$?
   (f) could f be continuous at $x = .5$ if $f(.5) = -0.75$?

6. If $1 < f(x) < 2$ for $0 < x < 1$ and f is continuous at $x = 0$, what can you say about $f(0)$? Be precise.

Answers to Continuity Probes

1. (a) 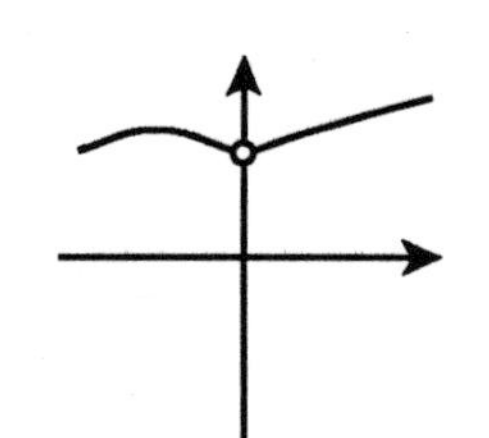

$f(a)$ not defined

(b) 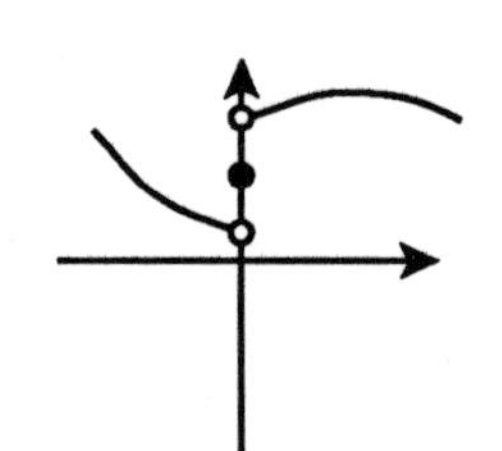

(c) 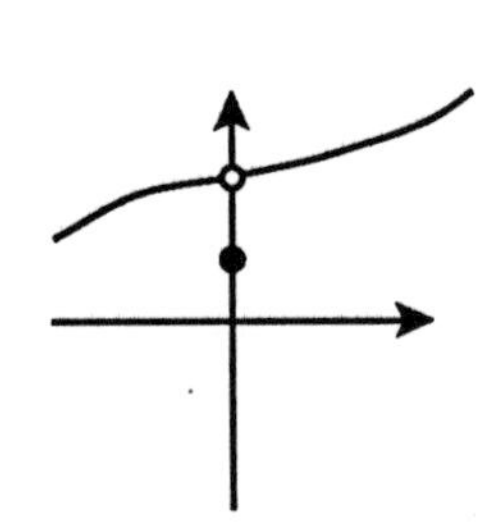

2. The function whose graph is shown in 1(c) is an example.

3. The function whose graph is shown in 1(b) is an example.

4. (a) Yes. For example, we might have $f(x) = 10^{11}x$ whose graph is a very steep line but is nevertheless continuous. Continuity at $x = a$ gives qualitative but not precise quantitative information about f "near" a. For example, if x is "near" a, then $f(x)$ is "near" $f(a)$, but "near" has to be determined function by function.

   (b) This is the rest of the story from part (a). Such a δ *is* required by the definition of continuity at $x = 0$.

5. (a) No, $\lim_{x \to 1} f(x) = f(1) = 0$ would be impossible.

   (b) Sure. The value of f at $x = 0$ has nothing to do with continuity at $x = 1$.

   (c) Yes.

   (d) Yes.

   (e) No, $\lim_{x \to .5} f(x) = f(.5) = -2$ would be impossible.

   (f) Yes.

6. (a) $1 \le f(0) \le 2$. In order to have $\lim_{x \to 0} f(x) = f(0)$ with $1 < f(x) < 2$ for $x \epsilon (0, 1)$, the limit at zero could only have the value somewhere between 1 and 2 *inclusive*.

Proof that if f and g are Both Continuous at $x = a$, so is $f + g$.

Question: Why is it important to know that sums, products, differences, and quotients (denominator non-zero) of continuous functions are themselves continuous?

Answer: It is easy to see that $f(x) = x$ is continuous at any point $x = a$ (because for a given ϵ we may choose $\delta =$ ____________). It is also easy to see that any constant function $g(x) = k$ is continuous (because for a given ϵ we may choose $\delta =$ ____________). From this we can conclude that functions like $h(x) = \dfrac{3x^2 - \pi x + 47}{40x^2 + 10}$ are continuous, not by giving an ϵ, δ argument, but because h is "built" from $f(x) = x$ and constant functions using only sums, products, differences, and quotients. Think how messy an ϵ, δ argument would be for such a complicated function as h. Then we can truly appreciate the importance of the theorem that tells us continuity is preserved by sums, products, differences, and quotients (where the denominator is non-zero).

Theorem: If f and g are continuous at $x = a$, then so is $f + g$.

Exercises on Continuity of $f + g$

1. If $f(x)$ differs from $f(a)$ by no more than $\frac{1}{10}$ for all $x \in (a-1, a+1)$, and $g(x)$ differs from $g(a)$ by no more than $\frac{1}{10}$ for all $x \in (a-1, a+1)$, then $f(x) + g(x)$ differs from $f(a) + g(a)$ by no more than ____________ for all $x \in$ (_______, _______).

   Draw an appropriate figure on the graph.

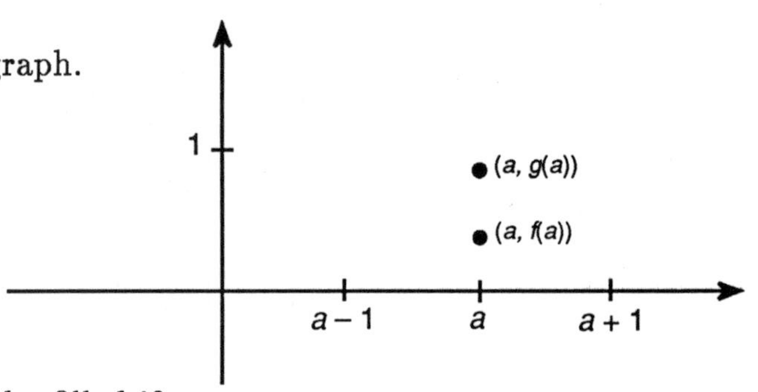

2. How would the blanks in problem 1 be filled if:

   (a) each $\frac{1}{10}$ was changed to $\frac{1}{100}$? The statement should now end "$\cdots$ by no more than ____________ for all $x \in$ (_______, _______)."

   (b) the first $\frac{1}{10}$ was changed to $\frac{1}{20}$? The statement should now end "$\cdots$ by no more than ____________ for all $x \in$ (_______, _______)."

   (c) each $\frac{1}{10}$ was changed to $\frac{\epsilon}{2}$? The statement should now end "$\cdots$ by no more than ____________ for all $x \in$ (_______, _______)."

   Comment: Exercises 1 and 2 show how a proof of the theorem above would go. If we must keep $| (f+g)(x) - (f+g)(a) | < \epsilon$ (to show $f+g$ is continuous at $x = a$), it suffices to make both $| f(x) - f(a) | < \frac{\epsilon}{2}$ and $| g(x) - g(a) | < \frac{\epsilon}{2}$. The latter can be done on some neighborhood of a (maybe not $(a-1, a+1)$, but on *some* neighborhood), since both f and g are continuous at $x = a$.

   The following exercises give more practice in the type of reasoning required in continuity proofs.

3. (a) If $f(x)$ differs from $f(a)$ by no more than $\frac{1}{10}$ for all $x \in (a-1, a+1)$, and $g(x)$ differs from $g(a)$ by no more than $\frac{1}{10}$ for all $x \in (a-1, a+1)$, then $f(x) - g(x)$ differs from $f(a) - g(a)$ by no more than ____________ for all $x \in$ (_______, _______).

   (b) Repeat (a) except replace each $\frac{1}{10}$ by $\frac{\epsilon}{2}$. The statement should now end "$\cdots$ by no more than ____________ for all $x \in$ (_______, _______)."

(c) Repeat (a) except replace "$x \in (a-1, a+1)$" by "$x \in (a-\frac{1}{2}, a+\frac{1}{2})$." The statement should now end "$\cdots$ by no more than ________ for all $x \in$ (____,____)."

4. If $f(x)$ differs from 2 by no more than $\frac{1}{2}$ for all $x \in (-1,1)$, and $g(x)$ differs from 3 by no more than $\frac{1}{2}$ for all $x \in (-1,1)$, then $f(x) \cdot g(x)$ differs from 6 by no more than ________ for all $x \in$ (____,____).
(Think: what is the largest and smallest product possible for $f \cdot g$ on $(-1,1)$?)

5. If $3 < f(x) < 4$ for all $x \in (-1,1)$, and $2 < g(x) < 3$ for all $x \in (-1,1)$, then

(a) ________ $< f(x) + g(x) <$ ________ for all $x \in$ ________.
(b) ________ $< f(x) \cdot g(x) <$ ________ for all $x \in$ ________.
(c) ________ $< f(x) - g(x) <$ ________ for all $x \in$ ________.
(d) ________ $< f(x)/g(x) <$ ________ for all $x \in$ ________.

Answers to Exercises on Continuity of $f+g$

Under "Answer:" for $f(x) = x$ take $\delta = \epsilon$
for $f(x) = k$ take δ *any* positive number.

1. $\frac{1}{10} + \frac{1}{10} = \frac{1}{5}$ for all $x \in (a-1, a+1)$. For the graph, drawing two lines, both with slope 1 is instructive. The sum $f+g$ then has slope 2.

2. (a) $\frac{1}{100} + \frac{1}{100} = \frac{1}{50}$ for all $x \in (a-1, a+1)$

   (b) $\frac{1}{20} + \frac{1}{10} = \frac{3}{20}$ for all $x \in (a-1, a+1)$

   (c) $\epsilon/2 + \epsilon/2 = \epsilon$ for all $x \in (a-1, a+1)$

3. (a) $\frac{1}{10} + \frac{1}{10}$ for all $x \in (a-1, a+1)$

   (b) $\epsilon/2 + \epsilon/2 = \epsilon$ for all $x \in (a-1, a+1)$

   (c) $\frac{1}{10} + \frac{1}{10}$ for all $x \in (a-1/2, a+1/2)$

4. The largest possible product is $5/2 \cdot 7/2 = 35/4$, which differs from 6 by $11/4$.

   The smallest possible product is $3/2 \cdot 5/2 = 15/4$, which differs from 6 by $9/4$.

   Thus, $f(x) \cdot g(x)$ differs from 6 by no more than $11/4$ for all $x \in (-1, 1)$.

5. (a) $3 + 2 = 5 \le f(x) + g(x) \le 4 + 3 = 7$ for all $x \in (-1, 1)$

   (b) $2 \cdot 3 = 6 \le f(x) \cdot g(x) \le 3 \cdot 4 = 12$ for all $x \in (-1, 1)$

   (c) $3 - 3 = 0 \le f(x) - g(x) \le 4 - 2 = 2$ for all $x \in (-1, 1)$

   (d) $3/3 = 1 \le f(x)/g(x) \le 4/2 = 2$ for all $x \in (-1, 1)$

THE INTERMEDIATE VALUE THEOREM

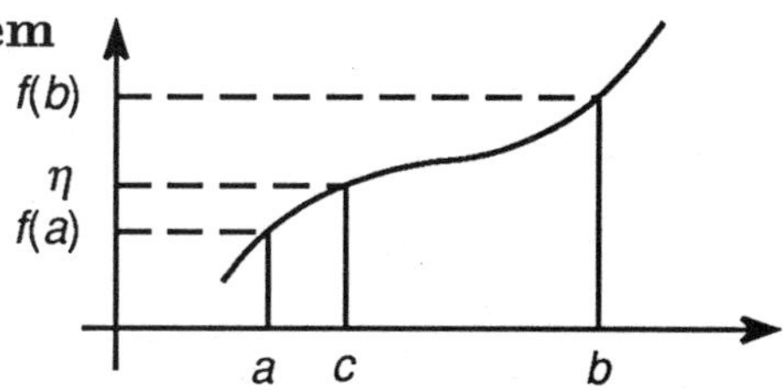

The Big Picture

The Intermediate Value Theorem

Theorem: If f is continuous on [a,b] and η is a number between $f(a)$ and $f(b)$ then for some $c \in [a, b]$, $f(c) = \eta$.

This theorem is geometrically obvious, so it is hardly a surprise. Our first reaction might be, "Why must we prove anything so obvious as that?" This "obvious" statement turns out to be *equivalent* to many less obvious statements such as the following (the list could be made much longer):

1. *Least Upper Bound (l.u.b.) Property of the Real Numbers.* Any non-empty set of real numbers that is bounded above has a least upper bound.
2. *Maximum Value Theorem.* Any continuous function on a closed bounded interval $[a, b]$ takes on a maximum value somewhere in the interval.
3. *Convergence of Bounded Monotone Sequences.* Any bounded monotone sequence converges. (If you haven't had sequences yet, don't worry; just realize a number of statements can be put in this list.)

The point is this: if we assume any of the statements in the list, then all of the others, along with the Intermediate Value Theorem, can be proved. Most textbooks assume the l.u.b. property and prove the others. If we chose to, we could take the statement of the Intermediate Value Theorem as an axiom and prove all these other important statements on our list.

The theorem is attributed to Bernardus Placidus Johann Nepomuk Bolzano (1781-1848), a Catholic priest with an impressive name from Bohemia, who was apparently the first to explicitly state the theorem. Even though the theorem is "obvious," it is used to prove some very surprising results that arise in a number of settings. The following pages give some examples.

Fixed Points and the Backpacker Problem

An Application of the Intermediate Value Theorem

Déja vu

A backpacker leaves the trailhead on Monday at 9 a.m. and walks a trail arriving at Hidden Lake at 3 p.m. His journey may have included rest stops, lunch, a nap, and returning to a stream crossing where fishing rods were left after getting a drink. On Tuesday, the backpacker leaves Hidden Lake at 9 a.m. and returns along the same trail. We aren't given the time of arrival at the trailhead.

Prove: The backpacker was at some location at the exact same time on both Monday and Tuesday.

One argument uses no calculus. Let a clone of the backpacker walk the Tuesday trek from Hidden Lake to the trailhead on Monday, while the backpacker walks Monday's trek from the trailhead to Hidden Lake. The two will meet somewhere along the way. The problem can be formulated in terms of *fixed points*, which play a central role in many mathematical settings and about which there is extensive literature. After reading the following theorem and proof, see if you can give a similar proof for the backpacker problem that uses the Intermediate Value Theorem.

Theorem: If f maps $[0,1]$ continuously into itself, then for some $x \in [0,1]$, $f(x) = x$ (that is, x is fixed under f).

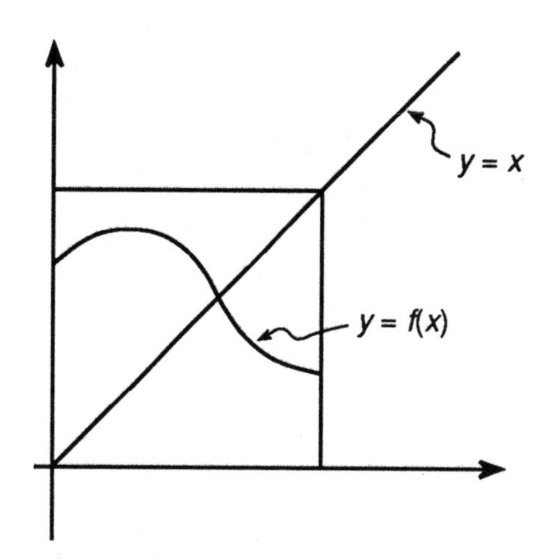

Proof: If $f(0) = 0$ or if $f(1) = 1$, we have our fixed point and are done. Otherwise, $f(0) > 0$ and $f(1) < 1$, since the range of f is a subset of $[0,1]$. Define $g(x) = f(x) - x$. Then g is continuous on $[0,1]$, $g(0) > 0$ while $g(1) < 0$. By the Intermediate Value Theorem, $g(x) = 0$ for some $x \in (0,1)$. Then $f(x) - x = 0$ and x is fixed under f.

Brouwer's Fixed Point Theorem generalizes the above theorem to n-dimensional space. In the plane it states that any continuous function of the unit disk $\{(x,y) : x^2 + y^2 \leq 1\}$ into itself leaves some point fixed. The disk can be replaced by other shapes. If a piece of paper is wadded up and placed where the original sheet lay, some point of the paper will be directly above its original position.

Proof For the Backpacker Problem. Let $d_1(t)$ be the distance (by trail) of the backpacker from the car on Monday. Let $d_2(t)$ be the distance of the backpacker from the car on Tuesday. We agree that $t = 0$ corresponds to 9 a.m.

Let $D(t) = d_1(t) - d_2(t)$. D is continuous and $D(0) < 0$ and $D(t_0) > 0$ where t_0 is a time of day that the backpacker is at the lake on Monday and the car on Tuesday. D is continuous on $[0, t_0]$. Now apply the Intermediate Value Theorem.

The Unstable Table

An Application of the Intermediate Value Theorem

Max bought a game table from Tablerite, Inc. for his basement, whose concrete floor was poured by Rock N. Roland Cement Works. Unfortunately, his table rocks back and forth on its four legs. Tablerite has guaranteed their laser construction places the bottoms of the four legs in a square that lies in a plane, and so Max is sure his problem is in a warped floor. What can he do?

Carol asks, "Don't you know the Intermediate Value Theorem for continuous functions on closed intervals? Simply rotate the table and it will stabilize in less than a 90° turn." After stabilizing the table, she showed Max the following proof.

Assume two opposite leg ends are squares and the other two are circles. Choose $\theta = 0$ as shown in the figure. Rotate the table through the angle θ and define $h_1(\theta)$ to be the combined heights of the square legs from the floor. Let $h_2(\theta)$ be the combined heights of round legs from the floor. Let $H(\theta) = h_1(\theta) - h_2(\theta)$. Assume that at $\theta = 0$ the table is unstable because not both round legs are on the floor. Then $h_1(0) = 0$, $h_2(0) > 0$ so that $H(0) < 0$. But at $\theta = \pi/2$ the legs change positions so $H(\pi/2) > 0$. Then for some $\theta_0 \in [0, \pi/2]$, $H(\theta_0) = 0$, that is, $h_1(\theta_0) = h_2(\theta_0)$, and since the table rests on three legs at all times, either $h_1(\theta_0)$or $h_2(\theta_0)$ is known to be zero. Hence $h_1(\theta_0) = h_2(\theta_0)$ and all four legs are on the floor.

Temperatures on the Equator

An Application of the Intermediate Value Theorem

We can prove that at any given instant there are two antipodal points on the equator with the same temperature.

Theorem: If the temperature on a circle varies continuously around the circle, then there are two antipodal points with the same temperature.

Proof:

Let $t(\theta)$ be the temperature on the circle at the point at angle θ on the circle. Define $T(\theta) = t(\theta) - t(\theta + \pi)$. Then T is continuous on $[0, \pi]$. If $T(0) = 0$, then $t(0) = t(0 + \pi) = t(\pi)$ and the antipodes with equal temperatures have been found. Otherwise, assume $T(0) > 0$. (The case $T(0) < 0$ is handled similarly.)
Then $T(0) = t(0) - t(\pi) > 0$ while $T(\pi) = t(\pi) - t(2\pi) = t(\pi) - t(0) < 0$. For some $\theta_0 \in [0, \pi], T(\theta_0) = 0$, that is $t(\theta_0) = t(\theta_0 + \pi)$.

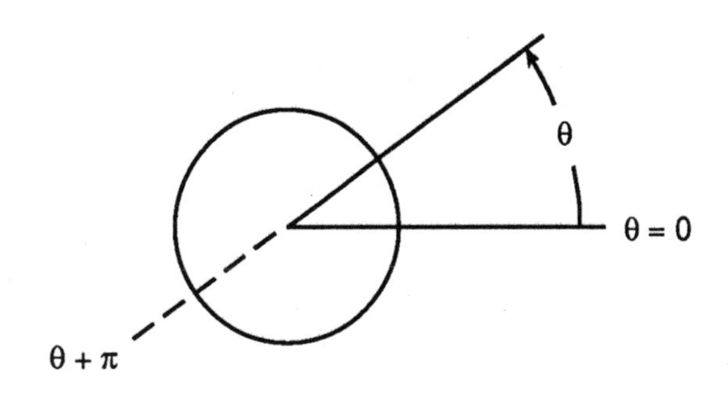

As you might expect, the theorem has analogues in higher dimensions. Borsuk's Theorem states that at any given time there are two antipodes on the surface of the earth that have the exact same temperature *and* barometric pressure. From the theorem above, we can say there is one antipodal pair *on the equator* with the same temperature, and some other antipodal pair *on the equator* with the same barometric pressure. Borsuk's Theorem tells us there is *one* antipodal pair somewhere on earth, not necessarily on the equator, where *both* temperature and barometric pressure agree. The proof of Borsuk's Theorem is much more difficult and requires essentially different tools in the proof.

Exercises on Cake Cutting

Applications of the Intermediate Value Theorem

1. 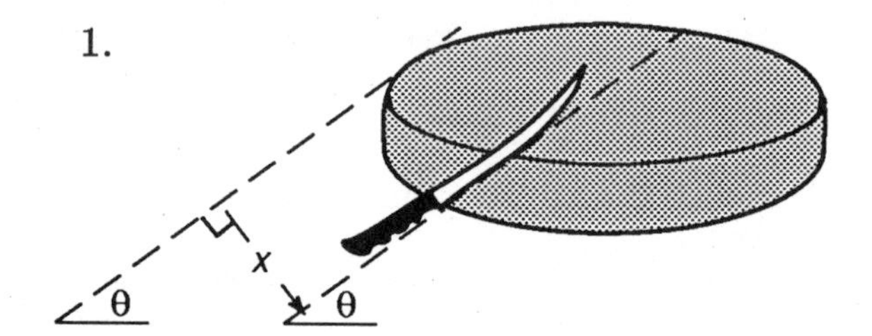

Show that by moving a knife parallel over a cake (regardless of shape) from an edge, there is a unique distance at which the knife will cut the cake in two pieces of equal area.

[*Hint*: Consider the function $L(x)$ = area of cake to the left of the knife, which can be assumed to be continuous.]

2. Suppose the cake is only partially frosted. Show that a single straight knife slice can simultaneously cut cake *and* frosting into equal areas.

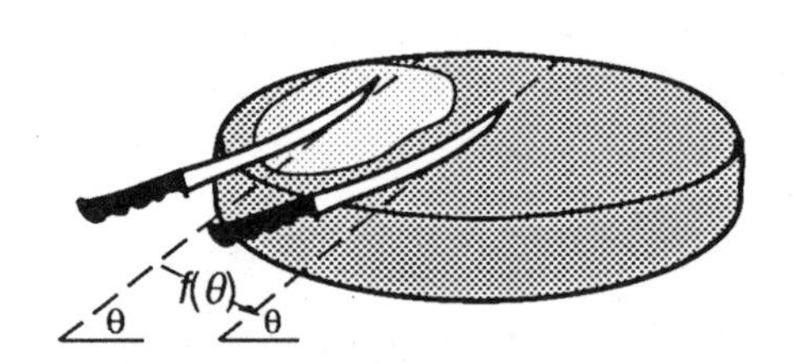

[*Hint*: Let $\theta, 0 \leq \theta \leq \pi$ be the cutting angle of two parallel cuts, one that cuts the area of the cake in half, the other that cuts the area of the frosting in half. Consider the function $f(\theta)$ = distance to the right from cake-bisecting cut to the frosting-bisecting cut. For the θ shown in the figure, $f(\theta) < 0$, since the knife bisecting the frosting is to the left of the cake-bisecting knife. f can be assumed to be continuous.]

3. Finish the following statement of the classical "Ham Sandwich Theorem," which generalizes the result above to three dimensions. "Suppose one has a sandwich made of bread, ham, and cheese. Then with one (planar) cut of a knife, ______________________________."

4. Show that the area of any cake with a convex boundary can be cut in equal thirds by two parallel cuts of equal length.

5. Show that the area of any cake with a convex boundary can be cut in equal fourths by two perpendicular cuts.

6. Prove that some cut simultaneously divides the area and perimeter of a convex cake in half.

Answers to Exercises on Cake Cutting

1. 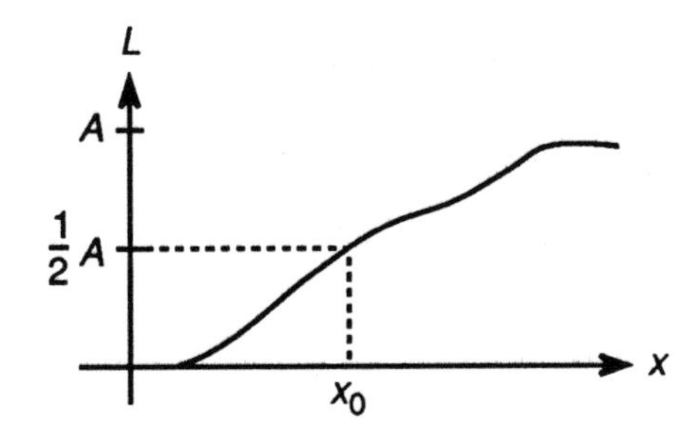

L is a continuous function that increases from 0 to A = area of cake. By the Intermediate Value Theorem, $L(x_0) = \frac{1}{2}A$ for some x_0. The x_0 is unique since L is increasing.

2. We require a value of θ where $f(\theta) = 0$. If $f(0) = 0$ we are done. If not, then $f(0)$ and $f(\pi)$ have opposite signs, and so there is some angle $\theta_0 \in (0, \pi)$ where $f(\theta_0) = 0$.

3. " ... all three ingredients can be simultaneously halved by the plane of the knife cut."

4. For any given θ, let $l_1(\theta)$ be the length of the segment that cuts a third off one end of the cake and $l_2(\theta)$ be the length of the segment that cuts a third off the other end. Let $L(\theta) = l_1(\theta) - l_2(\theta)$. If $L(0) = 0$ we are done. If not, $L(0)$ and $L(\pi)$ have opposite signs (the segments are directed), so $L(\theta_0) = 0$ for some $\theta_0 \in (0, \pi)$. That angle θ_0 is the direction of the required cuts.

5. For a given θ, let l_1 and l_2 be perpendicular lines that each cut the area in half as shown. Note that $S_1 + S_2 = .5, S_2 + S_3 = .5, S_3 + S_4 = .5$, and $S_1 + S_4 = .5$. Then if any of the four parts is a fourth, they all are. Suppose S_1 is too small, so S_4 is too big. By rotating through $90°, S_1$ becomes S_4–so somewhere in the process S_1 is exactly $\frac{1}{4}$.

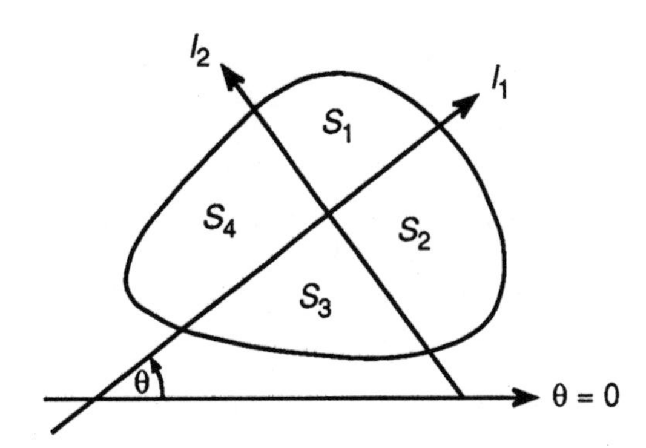

6. For an angle θ, let $l_1(\theta)$ cut the area in half and $l_2(\theta)$ (a parallel line) cut the perimeter in half. If $l_1(\theta)$ is to the left of $l_2(\theta)$ (assume that the lines are directed), then the opposite is true of $l_1(\theta + \pi)$ and $l_2(\theta + \pi)$. Hence, for some angle between θ and $\theta + \pi$ the two lines coincide.

Properties of Periodic Functions*

An Application of The Intermediate Value Theorem

Periodic functions are a familiar part of our world. Their graphs can be the familiar sine or cosine graphs or more complicated curves, such as the ones a doctor examines when reading an electrocardiograph. The curve shown is a computer-generated graph for $y = \sin x \cos 2x + \sin 4x$.

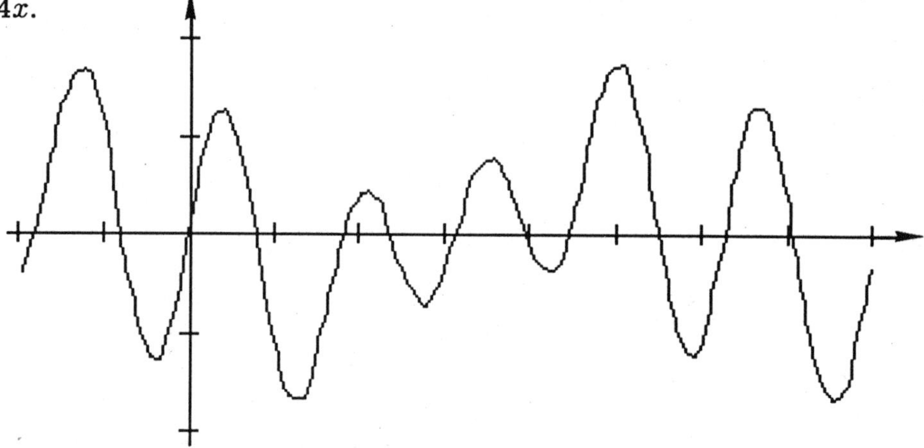

Regardless how "complicated" the periodic graph may be, certain properties are guaranteed. For example, given a straw of *any* length, it can be placed horizontally in some location so that both endpoints lie on the graph. Try this with some randomly-cut straw segments on the graph above!

Theorem: If f is continuous, has period p, and L is any length, there is a horizontal segment of length L which has both ends on the graph of f.

Proof: Since f has period p,

$$\int_0^p f(x)dx = \int_0^p f(x+L)dx.$$

Thus, $\int_0^p [f(x) - f(x+L)]dx = 0$. (#1. Why?) But the integrand is continuous and is therefore identically zero, or takes on both positive and negative values. (#2. Why?) By the Intermediate Value Theorem, for some $x_0 \in [0,p], f(x_0) - f(x_0 + L) = 0$. The desired segment joins $(x_0, f(x_0))$ and $(x_0 + L,\ f(x_0 + L))$ on the graph of f.

One can use the idea shown in the proof to get other interesting information about the graph of f. For example, $\int_0^p (f(x) + f(x+2L) - 2f(x+L))dx = 0$. For some $x_0 \in [0,p], f(x_0)+f(x_0+2L)-2f(x_0+L) = 0$; that is, $f(x_0+L) = [f(x_0)+f(x_0+2L)]/2$. This says the segment joining $(x_0, f(x_0))$ and $(x_0+2L, f(x_0+2L))$ is bisected by the point $(x_0 + L, f(x_0 + L))$on graph. So if f is periodic and continuous, given a segment of any

*This application requires some knowledge of integration to understand the proof of the theorem. Teachers may want to wait until students have studied integration before investigating this application.

length, it can be placed (not necessarily horizontally) so that its two endpoints and midpoints are all on the graph. Try this on the graph above, using a paper strip 2″ to 3″ long with a mark at its midpoint of an edge.

More generally, if $L_1 < L_2 < \cdots < L_n$ are *any* n given numbers then for some $x_0, f(x_0 + L_1) + \cdots + f(x_0 + L_n) = nf(x_0)$. Graphically interpreted, this says the y value at x_0 is the average of the values at $x_0 + L_1, \cdots, x_0 + L_n$. Amazing!

Reference: R. P. Boas, Jr., "A Primer of Real Functions," *Third Edition, Carus Mathematical Monograph No. 13*, Mathematical Association of America (1982).

Answers to Properties of Periodic Functions

1. $\int_0^P f(x) = \int_0^P f(x+L)dx$ because the integral is the same over any period. Subtraction then gives $\int_0^P (f(x) - f(x+L))dx = 0$.

2. If the integrand were always of one sign, positive or negative, then the integrand would be either positive or negative, but not zero. Thus, the integrand is either identically zero or takes on both positive and negative values.

THE DERIVATIVE

The Big Picture

The Derivative

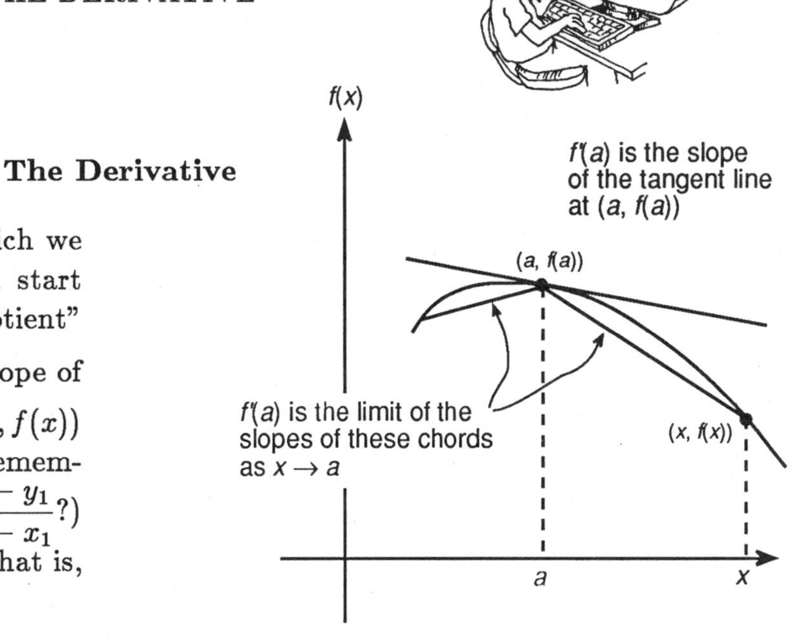

The derivative of f at $x = a$, which we denote by $f'(a)$, is a limit. You start with what is called a "difference quotient" $\frac{f(x) - f(a)}{x - a}$, which is simply the slope of the chord between the two points $(x, f(x))$ and $(a, f(a))$ on the graph of f. (Remember the formula for slope $m = \frac{y_2 - y_1}{x_2 - x_1}$?) Then we "allow x to approach a," that is, we take the limit, $\lim_{x \to a} \frac{f(x) - f(a)}{x - a}$. We see on our graph that when x is near a, the chords line up better with the tangent line to the curve at (a, f(a)), and so their slopes should approach as a limit the slope of the tangent line. When the limit of the slopes of these chords exists, that is our derivative at $x = a$. Be sure to have a good visual, intuitive grasp of what is happening when we write $f'(a) = \lim_{x \to a} \frac{f(x) - f(a)}{x - a}$.

The function has to be well-behaved near $x = a$ in order to have a derivative there. If f is not continuous at $x = a$, then the chords cannot line up with some non-vertical direction; that is, their slopes cannot approach a finite limit. So if $f'(a)$ exists f *must* be continuous at $x = a$. Also, if f is continuous at $x = a$ but the graph has a "corner" there, then the chords can't line up. Thus to have a derivative at $x = a$ requires continuity there, but functions can be continuous and still not have a derivative at a point. The derivative (the slope of the tangent line) will prove very useful.

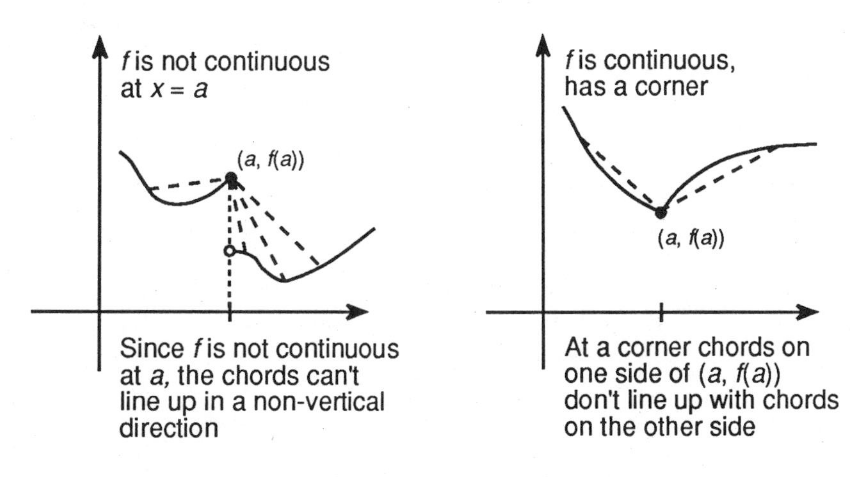

Note that anywhere the function has a maximum or minimum, the derivative (when it exists) must be zero, because tangent lines at a maximum or minimum on the curve will be horizontal lines. That simple geometric observation provides one of the most useful applications of the derivative that we will see. Another use of the derivative is to give a good approximation to your curve near $(a, f(a))$ with a straight line segment. The tangent line is that best linear approximation near $(a, f(a))$. Since we know its slope, $f'(a)$, and we have a point on the line, $(a, f(a))$, we can write the equation of that line using what we learned in our first algebra course:

$$y = f(a) + f'(a)(x - a).$$

In the days ahead, we will find out what a great idea this thing we call the derivative is!

Rules for Taking Derivatives

Sum: $(f+g)' = f' + g'$ Example: $\frac{d}{dx}(x + \sin x) = 1 + \cos x$

Difference: $(f-g)' = f' - g'$ Example: $\frac{d}{dx}(x^2 - x^3) = 2x - 3x^2$

Product: $(f \cdot g)' = f'g + fg'$ Example: $\frac{d}{dx}(x \cos x) = 1 \cdot \cos x + x(-\sin x) = \cos x - x \sin x$

Quotient: $(f/g)' = \dfrac{gf' - fg'}{g^2}$ Example: $\frac{d}{dx}\left(\frac{x}{1+x^2}\right) = \dfrac{(1+x^2) \cdot 1 - x(2x)}{(1+x^2)^2} = \dfrac{1-x^2}{(1+x^2)^2}$

Chain Rule: $(f(g))' = f'(g)g'$ Example: $\frac{d}{dx}(\sin x^2) = (\cos x^2) \cdot 2x$

Knowing the basic rules for differentiation (shown above) is a must. You will use them all the time, and so *the sooner you learn them well the better.* When you take further courses that use the calculus (Differential Equations will be the next one) you will repeatedly use *all* of these rules. None are optional or unimportant.

There is no surprise in the first two: the derivative of a sum or difference is the sum or difference of the derivatives. The same is *not* true of products or quotients. A nice rule would be that the derivative of a product is the product of the derivatives, but that isn't so. Don't join the multitude that fall into that trap! The reason the rules for products and quotients are what they are becomes apparent when you derive those rules–the algebra involved when working with the different quotient simply works out that way.

The Chain Rule may seem unnatural, but it is absolutely essential. If you know how to differentiate $\cos x$ and x^3, how do you differentiate $\cos x^3$? The Chain Rule tells you the answer to this very common question. The rule is written various ways. For example, if you are given a function $f(u)$, but u in turn is a function of x, then

$$\frac{d}{dx}f(u) = \frac{d}{du}f(u) \cdot \frac{d}{dx}(u) = \frac{df(u)}{du} \cdot \frac{du}{dx}.$$

Or if y is given in terms of u but u depends on x, then $\frac{dy}{dx} = \frac{dy}{du} \cdot \frac{du}{dx}$. You will become familiar with these various ways of saying the same thing. The important thing is to know how to apply the Chain Rule.

Besides these general rules, you have all the rules for specific functions such as $\frac{d}{dx}x^n = nx^{n-1}, n \neq 0$, or $\frac{d}{dx}\sin x = \cos x$, and so on. These also are important to know for a number of reasons. We use them often in applications, and when we get to integration (which reverses differentiation in a sense), you will be asking yourself questions such as: What function has as its derivative $x^2 + \sin x$? To answer a question like that you *must* know your rules for differentiation, so be forewarned and the wiser for it. *Learn your rules for differentiation thoroughly.* They won't go away.

An important skill to develop in differentiation is to recognize in which order the various rules apply. For example, suppose we are asked to find

$$\frac{d}{dx}\left[\left(x^2 \sin x^3 + (x^2+1)^3\right)\left(\frac{x}{1+e^{2x}}\right)\right].$$

Where do we start? What is the first rule to be applied? Answer: It is first a product, so we would write:

$$(x^2 \sin x^3 + (x^2+1)^3)'\left(\frac{x}{1+e^{2x}}\right) + (x^2 \sin x^3 + (x^2+1)^3)\left(\frac{x}{1+e^{2x}}\right)'.$$

Now we have two separate derivatives to take: the first is a sum, the second a quotient. This proceeds to unravel the sequence of steps required to compute the derivative. You will learn this by lots of practice. It can even be fun.

Using Rules for Differentiation

Recall some of our rules for differentiation.

1. $(f+g)' = f' + g'$
2. $(f-g)' = f' - g'$
3. $(f \cdot g)' = f' \cdot g + f \cdot g'$
4. $\left(\frac{f}{g}\right)' = \frac{g \cdot f' - f \cdot g'}{g^2}$
5. $(f(g))' = f'(g) \cdot g'$
6. $(x^n)' = nx^{n-1}$ if $n \neq 0$
7. $(\sin x)' = \cos x$
8. $(\cos x)' = -\sin x$
9. $(e^x)' = e^x$
10. $(\ln x)' = \frac{1}{x}$

1. To compute the derivatives of the following complicated functions, we would have to apply the above rules in a number of steps. But we must get started correctly. For each function, what is the *first* rule you would use in computing its derivative? Circle and draw an arrow pointing to where your attention is first focused. Then write the number of the first rule from page 37 that you would apply.

Function	*Rule*
Examples:	
$\frac{d}{dx}\left[\left(\sqrt{x}+e^x\right)\cdot\left(\sin x^3\right)\right]$	3
$\frac{d}{dx}\sqrt{\frac{x^2+1}{\sin x}}$	6
(a) $\frac{d}{dx}\left(\left(\sqrt{x}+e^x\right)\cdot \sin x^3\right)^4$	
(b) $\frac{d}{dx}\left(\frac{x+1}{x^2+1}\right)$	
(c) $\frac{d}{dx}\left(\ln\frac{x+1}{x^2+1}\right)^3$	
(d) $\frac{d}{dx}\frac{(x+1)^3}{\ln(x^2+1)}$	
(e) $\frac{d}{dx}\left(\frac{x^2+1}{x-1}-\sqrt{\sin x^2}\right)$	
(f) $\frac{d}{dx}\left(e^{\sin x^2}-x\right)$	
(g) $\frac{d}{dx}e^{\sin x^2-x}$	
(h) $\frac{d}{dx}\left[e^{\sin x^2-x}\cdot(\ln x+\cos x)\right]$	
(i) $\frac{d}{dx}\cos x^3$	

Function	*Rule*
(j) $\frac{d}{dx}\cos^3 x$	
(k) $\frac{d}{dx}\frac{\sin x}{\sin x^2}$	
(l) $\frac{d}{dx}\sin\left(\frac{x+1}{x-1}\right)^2$	
(m) $\frac{d}{dx}\sqrt{\sin\left(\frac{x+1}{x-1}\right)^2}$	
(n) $\frac{d}{dx}e^{6x^3}$	

Answers to Using Rules for Differentiation Probes

(a) 6

(b) 4

(c) 6

(d) 4

(e) 2

(f) 2

(g) 9

(h) 3

(i) 8

(j) 6

(k) 4

(l) 6

(m) 6

(n) 9

Exercises on Rules for Differentiation

1. Complete this table, which gives practice using the Chain Rule.

$f(g)$	f	g	$f'(g)$	g'	$\frac{d}{dx}f(g)$
Example: $\cos x^3$	cosine	cube	$-\sin x^3$	$3x^2$	$(-\sin x^3)(3x^2)$
(a) $\sqrt{\ln x}$					
(b) $\cos^3 x$					
(c) $e^{\sin x}$					
(d) $\ln x^2$					
(e) $(\ln x)^2$					

2. This is the same as exercise 2, but with more pizazz.

$f(g(h))$	f	g	h	$f'(g(h))$	$g'(h)$	h'	$\frac{d}{dx}f(g(h))$
Example: $\sqrt{\cos(x^{\frac{3}{2}})}$	sq. root	cosine	$\frac{3}{2}$-power	$\frac{1}{2}\left(\cos x^{\frac{3}{2}}\right)^{-\frac{1}{2}}$	$-\sin x^{\frac{3}{2}}$	$\frac{3}{2}x^{1/2}$	$\frac{1}{2}(\cos x^{3/2})^{-\frac{1}{2}}\cdot$ $(-\sin x^{3/2})(\frac{3}{2}x^{\frac{1}{2}})$
(a) $\cos^2 x^{3/2}$							
(b) $e^{\sin^2 x}$							
(c) $\cos^3 \sqrt{x}$							
(d) $\ln\cos^2 x$							
(e) $(\ln\cos x)^2$							
(f) $\ln\cos x^2$							

3. Each of the following is a derivative of a function obtained from the product rule. What was the function?

Derivative	Function
Example: $x^2 \sin x - 2x \cos x$	$-x^2 \cos x$
(a) $4x(x^2+1)\ln x + \dfrac{(x^2+1)^2}{x}$	
(b) $2 \sin x \cos x \sin x^2 + 2x \sin^2 x \cos x^2$	
(c) $\dfrac{3}{2}\sqrt{x}\ln(x^2+1) + \dfrac{2x^{5/2}}{x^2+1}$	
(d) $\dfrac{x \cos^3 x}{\sqrt{x^2+1}} - 3\sqrt{x^2+1}\cos^2 x \sin x$	
(e) $e^{x^2} + 2x^2 e^{x^2}$	

4. Each of the following is a derivative obtained by using the Chain Rule. Give a function with the given derivative.

Derivative	Function
Example: $3\cos^2 x(-\sin x)$	$\cos^3 x$
(a) $\dfrac{1}{2}(x^2 + \sin x)^{-\frac{1}{2}}(2x + \cos x)$	
(b) $\dfrac{1}{\sin x} \cdot \cos x$	
(c) $3e^x(e^x+1)^2$	
(d) $-\sin x^3(3x^2)$	
(e) $5(2x^2+\sqrt{x})^4\left(4x + \dfrac{1}{2\sqrt{x}}\right)$	

5. Each of the following is a derivative obtained by using the quotient rule. Give the function.

Derivative	Function
Example: $\dfrac{2x\cos x + x^2\sin x}{\cos^2 x}$	$\dfrac{x^2}{\cos x}$
(a) $\dfrac{2xe^x - (x^2+1)e^x}{e^{2x}}$	
(b) $\dfrac{\ln x - 1}{(\ln x)^2}$	
(c) $\dfrac{3x^2(x^2+1) - 2x^4}{(x^2+1)^2}$	
(d) $\dfrac{\sin x - x\cos x}{\sin^2 x}$	
(e) $\dfrac{1 - \ln x}{x^2}$	

Answers to Exercises on Rules for Differentiation

1. $f(g)$	f	g	$f'(g)$	g'	$\frac{d}{dx}f(g)$
(a) $\sqrt{\ln x}$	sq. root	log	$\frac{1}{2}(\ln x)^{-\frac{1}{2}}$	$\frac{1}{x}$	$\frac{1}{2}(\ln x)^{-\frac{1}{2}}(\frac{1}{x})$
(b) $\cos^3 x$	cube	cosine	$3\cos^2 x$	$(-\sin x)$	$(3\cos^2 x)(-\sin x)$
(c) $e^{\sin x}$	exp	sine	$e^{\sin x}$	$\cos x$	$e^{\sin x}(\cos x)$
(d) $\ln x^2$	log	square	$\frac{1}{x^2}$	$2x$	$(\frac{1}{x^2})2x$
(e) $(\ln x)^2$	square	log	$2(\ln x)$	$\frac{1}{x}$	$\frac{2}{x}\ln x$

2. $f(g(h))$	f	g	h	$f'(g(h))$	$g'(h)$	h'	$\frac{d}{dx}f(g(h))$
(a) $\cos^2 x^{3/2}$	square	cosine	$\frac{3}{2}$-power	$2\cos x^{\frac{3}{2}}$	$-\sin x^{\frac{3}{2}}$	$\frac{3}{2}x^{\frac{1}{2}}$	$(2\cos x^{\frac{3}{2}})(-\sin x^{\frac{3}{2}})(\frac{3}{2}x^{\frac{1}{2}})$
(b) $e^{\sin^2 x}$	exp	square	sine	$e^{\sin^2 x}$	$2\sin x$	$\cos x$	$(e^{\sin^2 x})(2\sin x)(\cos x)$
(c) $\cos^3\sqrt{x}$	cube	cosine	sq. root	$3\cos^2\sqrt{x}$	$-\sin\sqrt{x}$	$\frac{1}{2}x^{-\frac{1}{2}}$	$(3\cos^2\sqrt{x})(-\sin\sqrt{x})(\frac{1}{2}x^{-\frac{1}{2}})$
(d) $\ln\cos^2 x$	log	square	cosine	$\frac{1}{\cos^2 x}$	$2\cos x$	$-\sin x$	$(\frac{1}{\cos^2 x})(2\cos x)(-\sin x)$
(e) $(\ln\cos x)^2$	square	log	cosine	$2(\ln\cos x)$	$(\frac{1}{\cos x})$	$-\sin x$	$2(\ln\cos x)(\frac{1}{\cos x})(-\sin x)$
(f) $\ln\cos x^2$	log	cosine	square	$\frac{1}{\cos x^2}$	$-\sin x^2$	$2x$	$(\frac{1}{\cos x^2})(-\sin x^2)(2x)$

3. Derivative	Function
(a) $4x(x^2+1)\ln x + \dfrac{(x^2+1)^2}{x}$	$(x^2+1)^2 \ln x$
(b) $2\sin x \cos x \sin x^2 + 2x \sin^2 x \cos x^2$	$\sin^2 x \sin x^2$
(c) $\dfrac{3}{2}\sqrt{x}\ln(x^2+1) + \dfrac{2x^{5/2}}{x^2+1}$	$x^{\frac{3}{2}} \ln(x^2+1)$
(d) $\dfrac{x\cos^3 x}{\sqrt{x^2+1}} - 3\sqrt{x^2+1}\cos^2 x \sin x$	$\sqrt{x^2+1}\cos^3 x$
(e) $e^{x^2} + 2x^2 e^{x^2}$	xe^{x^2}

4. Derivative	Function
(a) $\dfrac{1}{2}(x^2+\sin x)^{-\frac{1}{2}}(2x+\cos x)$	$\sqrt{x^2+\sin x}$
(b) $\dfrac{1}{\sin x} \cdot \cos x$	$\ln \sin x$
(c) $3e^x(e^x+1)^2$	$(e^x+1)^3$
(d) $-\sin x^3(3x^2)$	$\cos x^3$
(e) $5(2x^2+\sqrt{x})^4 \left(4x + \dfrac{1}{2\sqrt{x}}\right)$	$(2x^2+\sqrt{x})^5$

5. Derivative	Function
(a) $\dfrac{2xe^x - (x^2+1)e^x}{e^{2x}}$	$\dfrac{x^2+1}{e^x}$
(b) $\dfrac{\ln x - 1}{(\ln x)^2}$	$\dfrac{x}{\ln x}$
(c) $\dfrac{3x^2(x^2+1) - 2x^4}{(x^2+1)^2}$	$\dfrac{x^3}{(x^2+1)}$
(d) $\dfrac{\sin x - x\cos x}{\sin^2 x}$	$\dfrac{x}{\sin x}$
(e) $\dfrac{1-\ln x}{x^2}$	$\dfrac{\ln x}{x}$

Derivatives and Differentials

If $y = f(x)$, we define the derivative of f at x to be

$$f'(x) = \lim_{h \to 0} \frac{f(x+h) - f(x)}{h}$$

when that limit exists. Like continuity, the derivative is defined *pointwise*. In fact, there are functions that have a derivative at exactly one point and no other. Luckily such functions are not met under reasonable circumstances, so they won't bother us. The point is, when we write $\frac{d}{dx}x^2 = 2x$ we are saying the function x^2 has a derivative *at each point* of its domain, and at a given but arbitrary point x the value of that derivative is $2x$.

There are a number of ways to write the derivative of f, including f', $\frac{df}{dx}$, and $\frac{dy}{dx}$ (if we have set $y = f(x)$). They all mean the same thing. The limit can be written in equivalent ways also, for example:

$$\lim_{h \to 0} \frac{f(x+h) - f(x)}{h} = \lim_{t \to x} \frac{f(t) - f(x)}{t - x}.$$

You need to have a good geometric understanding of all the various expressions used. Exercises have been prepared to give you practice in making these geometric interpretations. Below are verbal descriptions. For the moment, think of x as fixed.

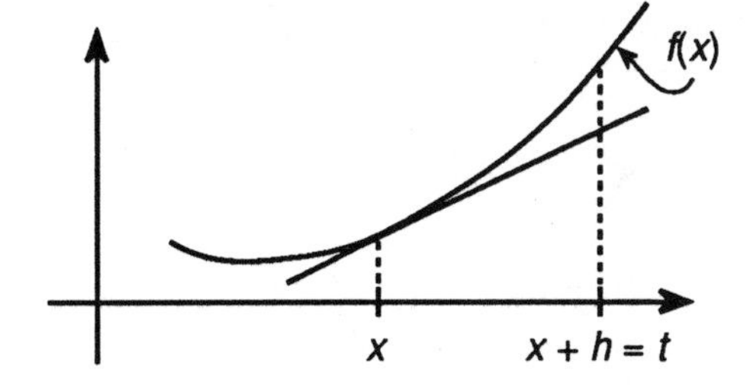

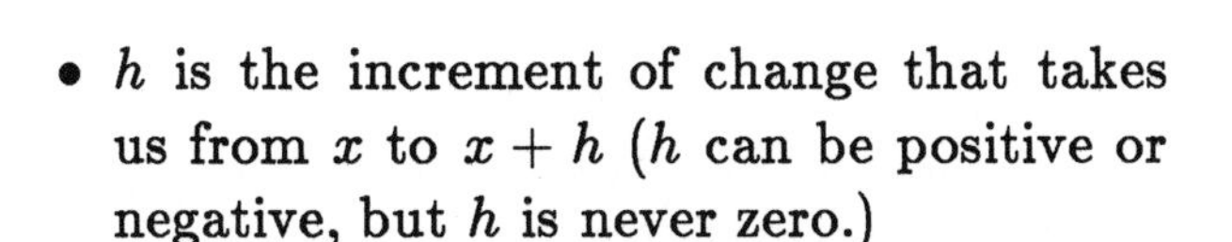

- h is the increment of change that takes us from x to $x + h$ (h can be positive or negative, but h is never zero.)
- $t - x$ plays the same role as h; it is the increment of change from the fixed point x to our point t that "moves" in the limit as $x + h$ "moves." We consider $h = t - x$ the increment in the independent variable, so $\Delta x = h = t - x$.
- The change Δx in the independent variable causes a corresponding change $\Delta y = f(x+h) - f(x) = f(t) - f(x)$ in the dependent variable. Like Δx, Δy may be positive or negative. Unlike Δx, Δy can be zero.
- $\dfrac{\Delta y}{\Delta x} = \dfrac{f(x+h) - f(x)}{h} = \dfrac{f(t) - f(x)}{t - x}$ is the slope of the line segment joining the two points $(x, f(x))$ and $(x+h, f(x+h)) = (t, f(t))$. It may look worse, but it is only the expression for the slope you learned in Algebra I. Don't let new notation make you think things are more complicated than they really are.

- The thing that is entirely new is the limit, $\lim_{h\to 0}\frac{f(x+h)-f(x)}{h}$. It has to be evaluated using the limit definition we have already seen. When the limit exists it will give a *number* that we denote by $f'(x)$. It has geometric meaning, as you have seen: it is what we define as the slope of the line tangent to the graph of f at $(x, f(x))$.

Be careful with notation. For the moment, suppose $f(x) = x^2$. Many students will incorrectly write things like

$$f'(x) = \lim_{h\to 0}\frac{(x+h)^2 - x^2}{h} = \frac{x^2 + 2xh + h^2 - x^2}{h} = \frac{h(2x+h)}{h} = 2x.$$

Notice that "$\lim_{h\to 0}$" should appear before the two quotients where it is left off. That symbol, "$\lim_{h\to 0}$," instructs us to take a limit, and cannot be dropped until we have done what it tells us to do. Notice also that $f'(x)$ and $\frac{f(x+h)-f(x)}{h}$ are two different things. One is the slope of a tangent line, and the other is a slope of a secant on the graph of f. Let's try it again and get it correct. If $y = f(x) = x^2$, then

$$\frac{dy}{dx} = \frac{df}{dx} = f'(x) = \lim_{h\to 0}\frac{(x+h)^2 - x^2}{h} = \lim_{h\to 0}\frac{x^2 + 2xh + h^2 - x^2}{h} = \lim_{h\to 0}\frac{h(2x+h)}{h} = 2x$$

The equation $f'(x) = \frac{dy}{dx}$ looks a bit strange; one side looks like a fraction, while the other doesn't. Is $\frac{dy}{dx}$ a fraction as we have known fractions all our life? Certainly $\frac{\Delta y}{\Delta x} = \frac{f(x+h)-f(x)}{h}$ is a fraction because it has a numerical numerator and denominator. But what about $\frac{dy}{dx} = \lim_{h\to 0}\frac{\Delta y}{\Delta x}$? After the limit is taken do we still have something that makes sense as a fraction? If $\frac{dy}{dx}$ is a "normal" fraction, both dy and dx should be numbers. What are they?

This brings us to *differentials*. They allow us to honestly treat $\frac{dy}{dx}$ as a fraction, and they will prove useful in making approximations. Start with

$$\frac{dy}{dx} = f'(x).$$

Let us *define* $dx = \Delta x$ to be any non-zero number you wish. We will think of it as an increment in the independent variable x. Now that gives us a numerical interpretation for dx. The rest is all set because we now know what $f'(x)$ and dx are. If $\frac{dy}{dx}$ is to be a fraction, we can apply the usual rules of algebra to solve for dy, finding $dy = f'(x)dx = f'(x)\Delta x$.

There is a nice geometric interpretation for dy. We know the slope of the line tangent to the curve at $(x, f(x))$ is $f'(x)$. From what we know from algebra about slopes, $m = \frac{\text{rise}}{\text{run}} = \frac{y_2 - y_1}{x_2 - x_1}$, where (x_1, y_1) and (x_2, y_2) are two points on our line. That means (slope of line)$\cdot$ run = rise. That makes sense. If we move along a line with slope 2 for a horizontal run of 3 units, we get a vertical rise of 6 units. Thus dy is the change in y that we get *if we move along the tangent line* at $(x, f(x))$ for a run of $dx = \Delta x$. On the other hand, $\Delta y = f(x+h) - f(x)$ is the change in y when *we move on the graph of* f from x for a distance Δx. In our figure, on page 47, $dy < \Delta y$, but for small $h = dx = \Delta x$, dy should be a good approximation to Δy because the tangent line "fits" the curve well for a while.

Let's summarize:

dx denotes an *increment*, or change, in domain values from x to $x + dx$. It is written interchangeably as
$dx = \triangle x = h$

$\triangle y$ denotes the *exact difference* in function values corresponding to the increment $\triangle x$:
$\triangle y = f(x + \triangle x) - f(x)$

dy denotes the *approximate change* in function values, as given by the change in y values along the line tangent to the graph of f at $(x, f(x))$:
$dy = f'(x)dx$

An important question remains: what use is there for the differential? To answer this, let us suppose $f(x_0)$ and $f'(x_0)$ are known, or are at least easy to compute, at the point x_0. We then have the approximate equation

$$f(x_0 + dx) \approx f(x_0) + dy,$$

or equivalently

$$f(x_0 + dx) \approx f(x_0) + f'(x_0)dx.$$

The exact function value $f(x_0 + dx)$ may be difficult to compute, but the approximate value $f(x_0) + f'(x_0)dx$ is simply a multiplication and an addition–could anything be easier?!

To illustrate this important idea, let's use the differential to approximate $\sqrt[3]{1030}$.

1. What function should we use? Choose $f(x) = \sqrt[3]{x} = x^{1/3}$, whose derivative is $(1/3)x^{-2/3}$.

2. Since we are trying to get an approximation of $\sqrt[3]{1030}$, we must choose $x_0 + dx = 1030$. What is x_0? We need a number x_0 where we can easily compute $f(x_0)$ and $f'(x_0)$. Choose $x_0 = 1000$, in which case $dx = 30$.

3. That makes $f(x_0) = \sqrt[3]{1000} = 10$ and $dy = f'(x)dx = \left((1/3)(1000)^{-2/3}\right) \cdot 30 = \left(1/3 \cdot \frac{1}{100}\right) \cdot 30 = \frac{1}{300} \cdot 30 = \frac{1}{10}$.

4. Thus, $\sqrt[3]{1030} \approx \sqrt[3]{1000} + dy = 10.1$.

You may be worried that $dx = 30$ is so large that your approximation is way off the mark. You may be surprised when you check how well you did by using a calculator to compute $\sqrt[3]{1030}$. You will discover that the approximation by the differential is in error by less than 0.01%!

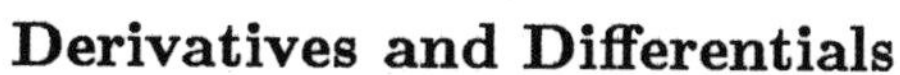

Derivatives and Differentials

1. Use the figure to match expressions on the right with the geometric interpretations on the left.

(i)	The slope of line tangent to curve at A	(a)	$\Delta x = x - a$
(ii)	Length of segment BD	(b)	$\Delta y = f(x) - f(a)$
(iii)	The slope of line tangent to curve at C	(c)	$dy = f'(a)(x - a)$
(iv)	Length of segment BC	(d)	$\lim\limits_{a \to x} \dfrac{f(x) - f(a)}{x - a}$
(v)	Length of segment AB	(e)	$\dfrac{f(x) - f(a)}{x - a}$
(vi)	Slope of segment AC	(f)	$(f(x) - f(a)) - f'(a)(x - a) = \Delta y - dy$
(vii)	Length of segment CD	(g)	$\lim\limits_{x \to a} \dfrac{f(x) - f(a)}{x - a}$
(viii)	Length of segment AD	(h)	$(x - a)\sqrt{1 + (f'(a))^2}$

2. Suppose $\Delta y = f(x) - f(a)$, $\Delta x = x - a$, $dy = f'(a)\Delta x$. Assume $x > a$.

(i) If $f'(x)$ is the same for each x (that is, the graph of f is a straight line), then:

(a) $\Delta y > dy$ (b) $\Delta y = dy$ (c) $\Delta y < dy$ (d) can't say

(ii) If $f''(x) > 0$ for all x (that is, the graph of f is concave upward), then:

(a) $\Delta y > dy$ (b) $\Delta y = dy$ (c) $\Delta y < dy$ (d) can't say

(iii) If $f'(x) < 0$ for all x (that is, f is decreasing), then:

(a) $\Delta y > dy$ (b) $\Delta y = dy$ (c) $\Delta y < dy$ (d) can't say

Answers to Derivatives and Differentials Probes

1. i-g, ii-c, iii-d, iv-b, v-a, vi-e, vii-f, viii-h

2. i-b.

   ii-a, as in the figure, the approximation dy for the change in f undershoots the actual change Δy.

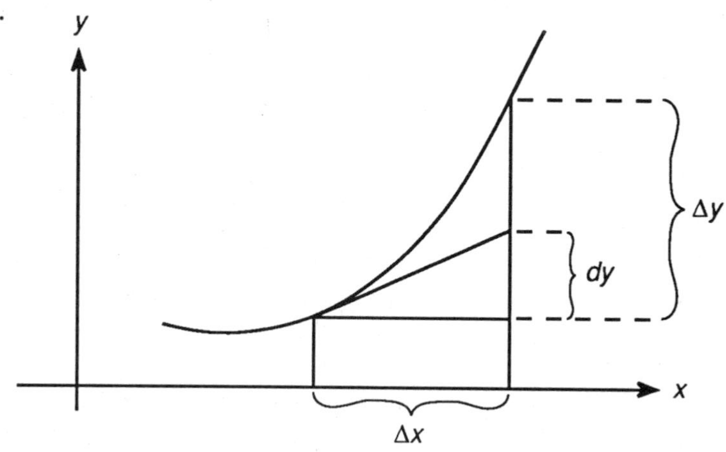

   iii-d. $f' < 0$ just means f is decreasing. f can still be either concave upward or downward, as the following two graphs show.

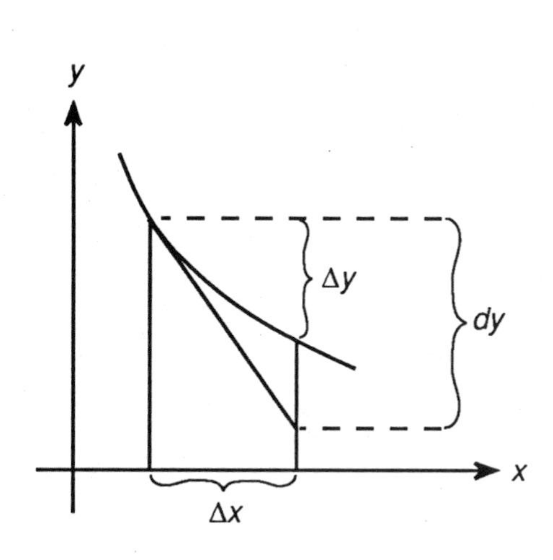

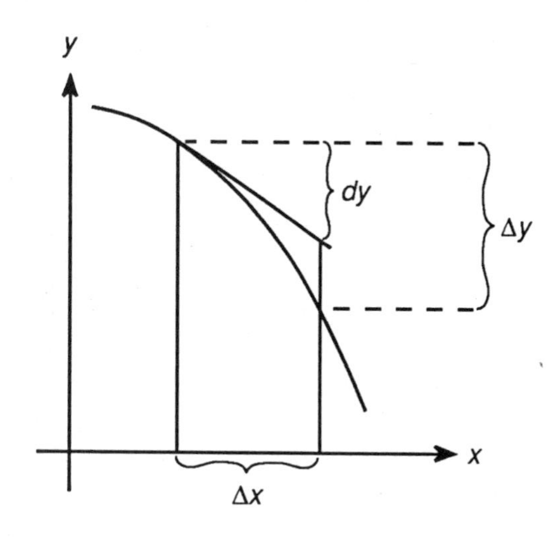

Exercises on Estimates Using the Derivative

1. If $f(a) = 10$ and $f'(a) = 2$, then the most reasonable guess for $f(a + \frac{1}{10})$ is ____________, and the most reasonable guess for $f(a - \frac{1}{3})$ is ____________.

2. If $f(\frac{1}{10}) = 3$ and $f(0) = 2$, then the most reasonable guess for $f'(0)$ is ____________.

3. If $f(\frac{1}{10}) = 3$ and $f(\frac{-1}{10}) = 2$ then the most reasonable guess for $f(0)$ is ____________ and $f'(0)$ is ____________.

4. If f is everywhere differentiable and $f(0) = 1$ and $f(3) = -6$, you can be sure that at some point $x = c$, $f'(c) =$ ____________. In fact, we can say $c \in$ (______, ______).

5. If $f(1) = 2, f'(1) = 3, G(2) = 7$, and $G'(2) = -3$, then the best estimate for $f(1.1)$ is ____________ and the best estimate for $G(f(1.1))$ is ____________.

6. If $f(1) = 3, f'(1) = 2, g(1) = 4, g'(1) = -1$, then the best estimate of

   (a) $f(.9) + g(.9)$ is ____________.

   (b) $f(1.1) - g(1.1)$ is ____________.

   (c) $f(.9) \cdot g(.9)$ is ____________.

   (d) $\dfrac{f(1.1)}{g(1.1)}$ is ____________.

Answers to Exercises on Estimates Using the Derivative

1. $f(a+\frac{1}{10}) \approx f(a) + f'(a)\left[(a+\frac{1}{10}) - a\right] = 10 + 2(\frac{1}{10}) = 10.2$
   $f(a-\frac{1}{3}) \approx f(a) + f'(a)\left[(a-\frac{1}{3}) - a\right] = 10 + 2(-\frac{1}{3}) = 9\frac{1}{3}$

2. $f'(0) \approx \dfrac{f(\frac{1}{10}) - f(0)}{\frac{1}{10} - 0} = \dfrac{3-2}{\frac{1}{10}} = 10$

3. $f(0) \approx \dfrac{f(\frac{1}{10}) + f(-\frac{1}{10})}{2} = \dfrac{3+2}{2} = \dfrac{5}{2}$
   $f'(0) \approx \dfrac{f(\frac{1}{10}) - f(-\frac{1}{10})}{\frac{1}{10} - (-\frac{1}{10})} = \dfrac{3-2}{\frac{2}{10}} = 5$

4. There is some $c \in (0,3)$ where $f'(c) = \dfrac{f(3) - f(0)}{3-0} = \dfrac{-6-1}{3-0} = \dfrac{-7}{3}$.

5. $f(1.1) \approx f(1) + f'(1)((1.1) - 1) = 2 + 3(.1) = 2.3$
   $G(f(1.1)) \approx G(f(1)) + G'(f(1))(f(1.1) - f(1)) = G(2) + G'(2)(f(1.1) - f(1)) \approx 7 + (-3)[2.3 - 2] = 7 - .9 = 6.1$

6. (a) $(f+g)(.9) \approx (f+g)(1) + [(f+g)'(1)]\,[.9-1] = f(1) + g(1) + (f'(1) + g'(1))(-.1) = 3 + 4 + (2 + (-1))(-.1) = 7 - (.1) = 6.9$

   (b) $(f-g)(1.1) \approx (f-g)(1) + [(f-g)'(1)]\,[1.1-1] = f(1) - g(1) + (f'(1) - g'(1))(.1) = 3 - 4 + (2 - (-1))(.1) = -1 + .3 = -.7$

   (c) $(f\cdot g)(.9) \approx (f\cdot g)(1) + (f\cdot g)'(1)(.9-1) = f(1)\cdot g(1) + [f(1)g'(1) + f'(1)g(1)]\,(-.1) = 3\cdot 4 + [3(-1) + 2\cdot 4]\,(-.1) = 12 + 5(-.1) = 11.5$

   (d) $(\frac{f}{g})(1.1) \approx (\frac{f}{g})(1) + (\frac{f}{g})'(1)(1.1-1) = \frac{f(1)}{g(1)} + \dfrac{g(1)\cdot f'(1) - f(1)g'(1)}{(g(1))^2}(.1)$
   $= \dfrac{3}{4} + \dfrac{4\cdot 2 - 3(-1)}{16}(.1) = \dfrac{3}{4} + \dfrac{11}{16}(.1) \approx .75 + .07 = .82$

Position, Velocity, and Acceleration

1. Given the graph of $v(t)$ for a particle moving along a straight line, draw a plausible graph for the corresponding functions $a(t)$ and $s(t)$. Assume $s(0) = 0$.

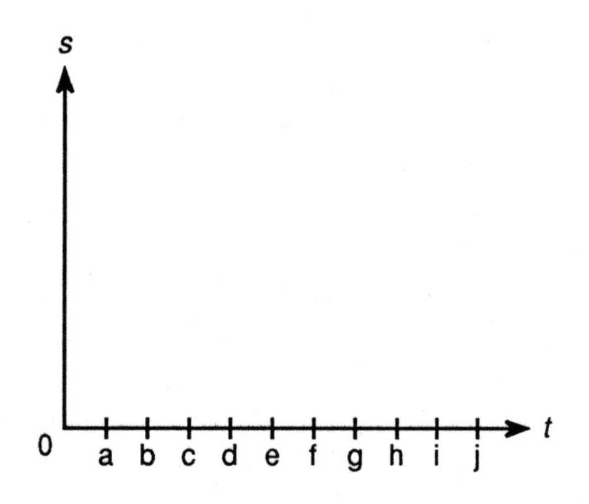

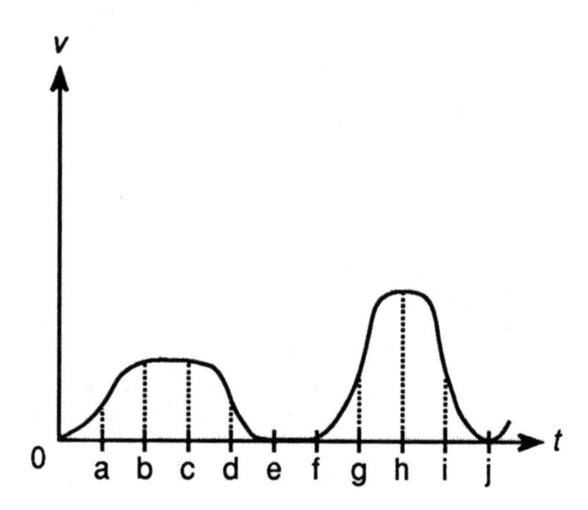

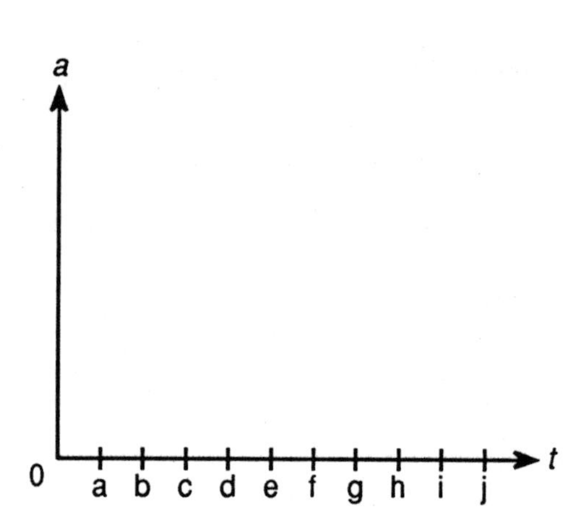

2. Find the intervals on which:

a.	$a(t) > 0$	e.	$s''(t) > 0$
b.	$a(t) = 0$	f.	$v'(t) > 0$
c.	$s(t) \geq 0$	g.	$a'(t) > 0$
d.	$s'(t) = 0$	h.	$a'(t) = 0$

Answers to Position, Velocity, and Acceleration Probes

1. 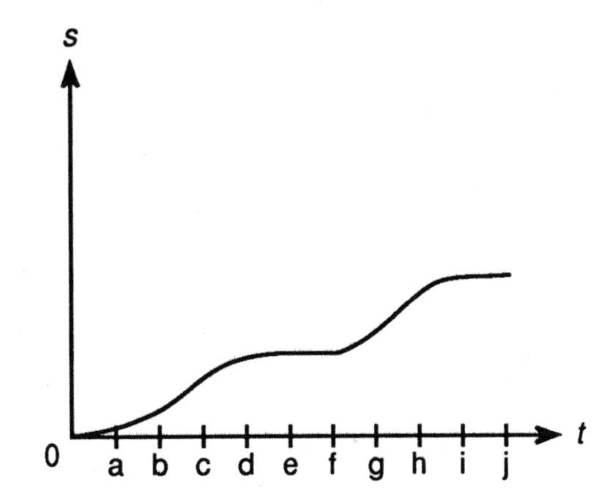

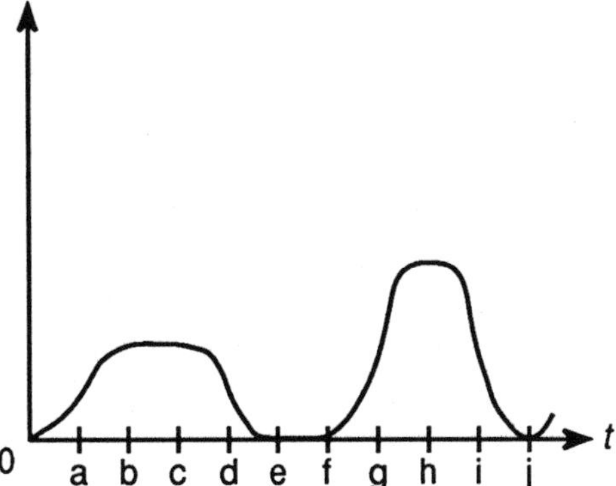

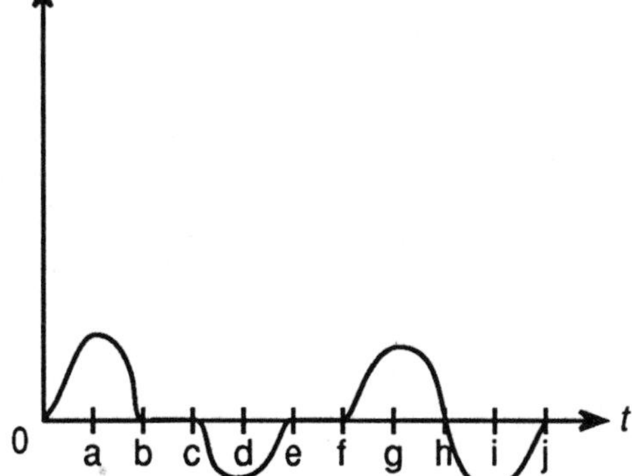

2. a. Where $v'(t) > 0$; that is, on $(0, b)$ and (f, h).

   b. Where $v'(t) = 0$; that is, on $[b, c], [e, f]$ and at $t = h$ and $t = j$.

   c. $s(t) \geq 0$ always, since $s(0) = 0$ and $v(t)$ is never negative.

   d. $s'(t) = v(t)$ so $s'(t) = 0$ at $0, j$, and on $[e, f]$.

   e. $s''(t) = a(t)$ so $s''(t) > 0$ on $(0, b)$ and (f, h). This is the same question as (a.) since $v' = s''$.

   f. $v'(t) = a(t)$, same as (e.) and (a.).

   g. Where $v(t)$ is concave upward since $a' = v''$; $(0, a), (d, e), (f, g)$, and (i, j).

   h. At inflection points on graph of v and where v is constant; therefore, at a, d, g, i, and on $[b, c]$ and $[e, f]$.

Comparing a^b and b^a

Which is larger, a^b or b^a?

As we shall see, the solution to the problem stated above is quite easy when one has graphing skills and a knowledge of L'Hôpital's Rule.

First of all, to compare a^b and b^a (a and b are assumed to be positive constants) we can take logarithms of both terms and see that it is enough to compare $b \ln a$ with $a \ln b$. Or, dividing both of these quantities by ab, we see that it is enough to compare $\frac{1}{a} \ln a$ with $\frac{1}{b} \ln b$. This suggests that we consider the function $f(x) = \frac{1}{x} \ln x$. Clearly $\lim_{x \to 0+} f(x) = -\infty$, and (by L'Hôpital's Rule) $\lim_{x \to \infty} f(x) = 0$. Since $f'(x) = \frac{1}{x^2}(1 - \ln x)$, we see $f'(x) > 0$ for $0 < x < e$ and $f'(x) < 0$ for $x > e$. Thus f has a maximum at $x = e$, where $f(e) = \frac{1}{e}$.

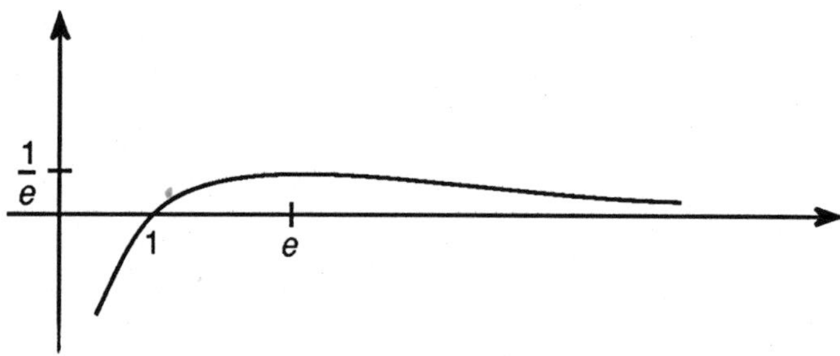

Since f increases on $(0, e]$, we see that $a^b < b^a$ when $0 < a < b \le e$. Similarly $a^b > b^a$ when $e \le a < b$ since f decreases on $[e, \infty)$.

Exercises on Comparing a^b and b^a

1. If $a^b = b^a$ and $a < b$, what else can you say about a and b?

2. Decide which (if either) of the following numbers is bigger:

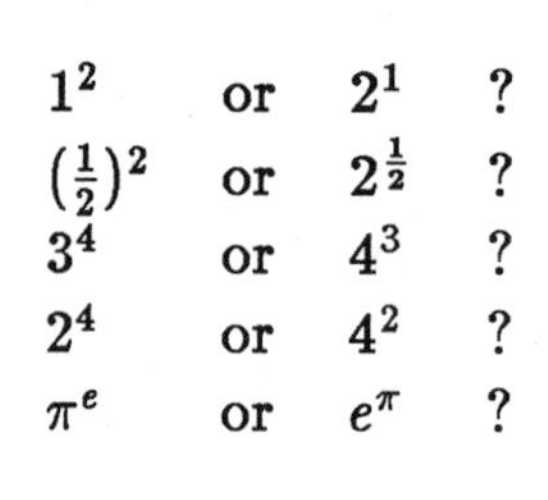

1^2	or	2^1	?
$(\frac{1}{2})^2$	or	$2^{\frac{1}{2}}$	?
3^4	or	4^3	?
2^4	or	4^2	?
π^e	or	e^π	?

Answers to Exercises on Comparing a^b and b^a

1. $a \in (1, e)$ $b \in (e, \infty)$

2. $1^2 < 2^1$, $(\frac{1}{2})^2 < 2^{\frac{1}{2}}$, $3^4 > 4^3$ (since $3 > e$), $2^4 = 4^2 = 16$, $\pi^e < e^\pi$ (since $e < \pi$)

Graphing with Information about the Derivatives

When possible, draw a graph of function f that satisfies all of the properties listed.

1. (a) $f(0) = 0$
   (b) $f(x) < 0$ for all $x > 0$
   (c) $f'(x) < 0$ for all $x > 0$
   (d) $f''(x) > 0$ for all $x > 0$

2. Add the condition $(e) f'''(x) > 0$ to each of the conditions in 1., above.

3. (a) $f(x) > 0$ for all $x > 0$
   (b) $f(x) < 0$ for all $x < 0$
   (c) $f(0) = 0$
   (d) $f'(x) > 0$ for all x
   (e) $f''(x) > 0$ for all $x < 0$
   (f) $f''(x) < 0$ for all $x > 0$

4. (a) $f(0) = 0$
   (b) $f(x) < 0$ for all $x > 0$
   (c) $f''(x) > 0$ for all $x > 0$
   (d) $\lim_{x \to \infty} f(x) = -\infty$

5. (a) $f(x) > 0$ for all x
   (b) $f''(x) > 0$ for all x
   (c) $f'''(x) = 0$ for all x

6. (a) $f(x) \neq 0$ for all x
   (b) $f'(x) > 0$ for all x
   (c) $f'''(x) = 0$ for all x

Answers to Graphing with Information about the Derivatives Probes

1.

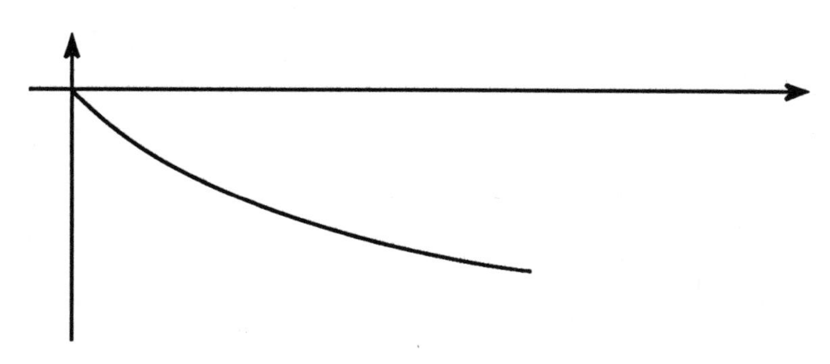

2. This can't be done. $f''' > 0$ says f'' is increasing, which graphically requires the curve to be more severely concave upward as we move right. You can't do that and satisfy $f(x) < 0$ for all $x > 0$.
3.

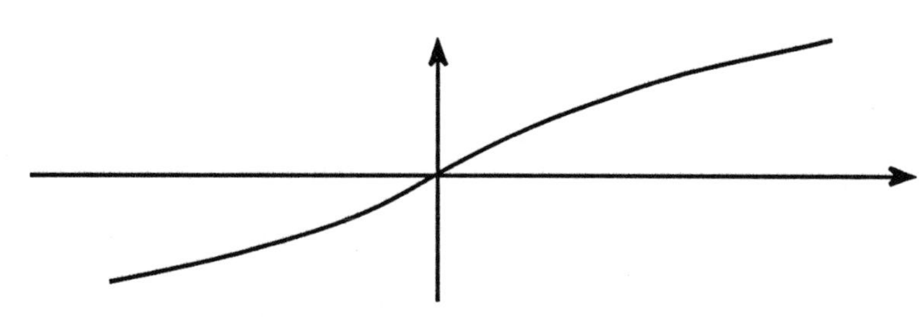

4. $f(x) = -\sqrt{x}$, for example.

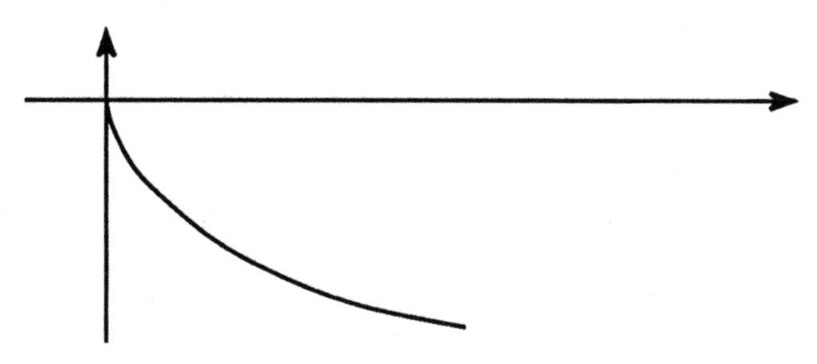

5. The graph must be an upward-opening parabola lying in the upper half-plane.

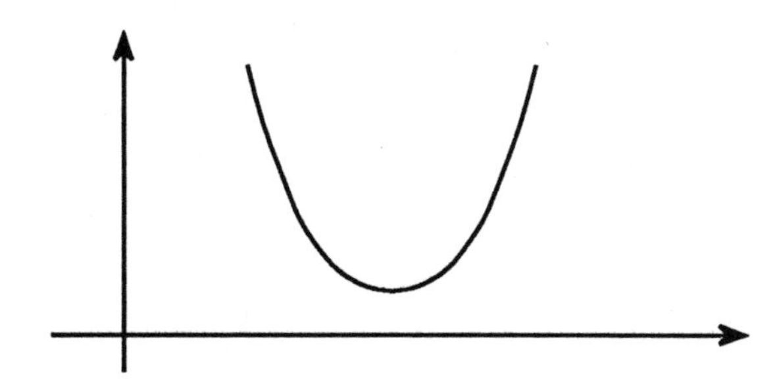

6. f'' must be constant so f' must have the form $ax + b$. But f' would have a zero unless $a = 0$, and so $f'(x) = b$ where $b > 0$. Then $f(x) = bx + c$, which has a zero. Thus, no function f exists that satisfies all three conditions.

Exercises on Relating the Graphs of Functions and their Derivatives

Each of the following columns is headed by the graph of a function f. Underneath this graph are three others that are, in random order, the graphs of f', f'' and a function F whose derivative is f; i.e., $F' = f$. Label each graph with the correct function.

1.

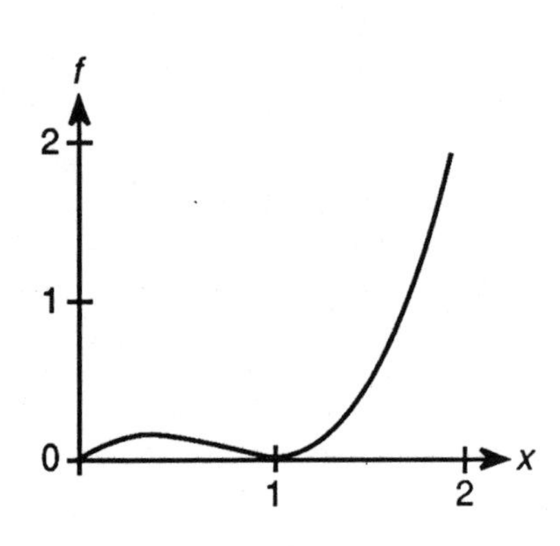

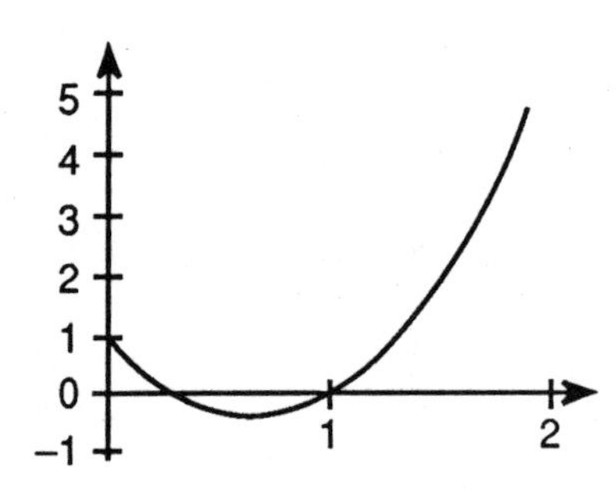

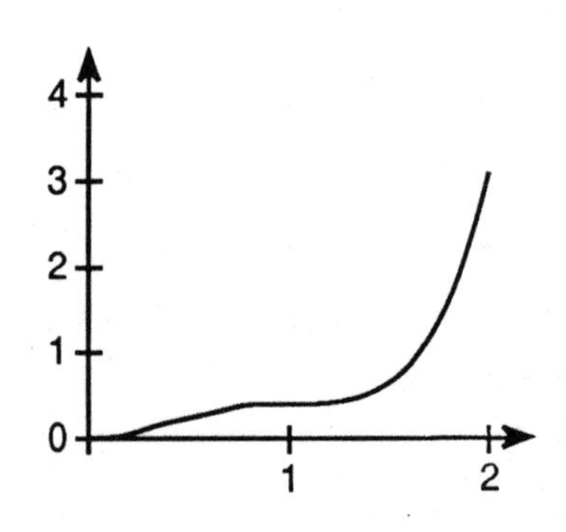

2.

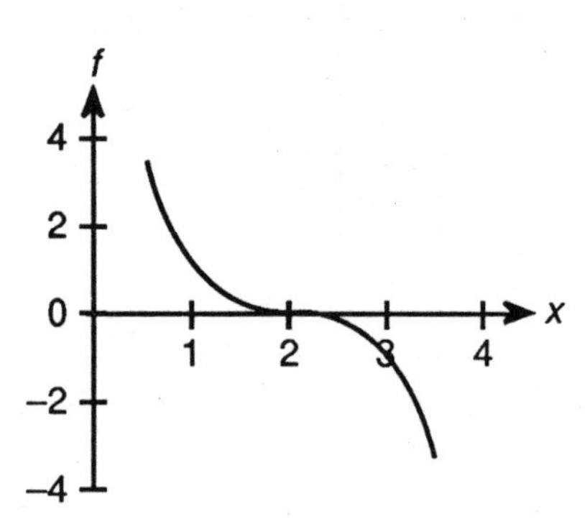

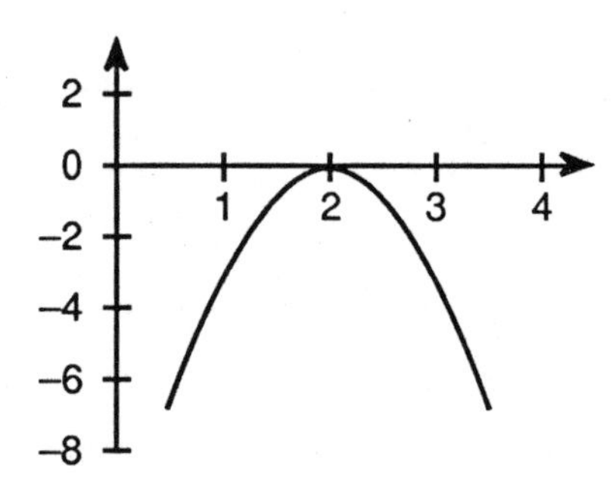

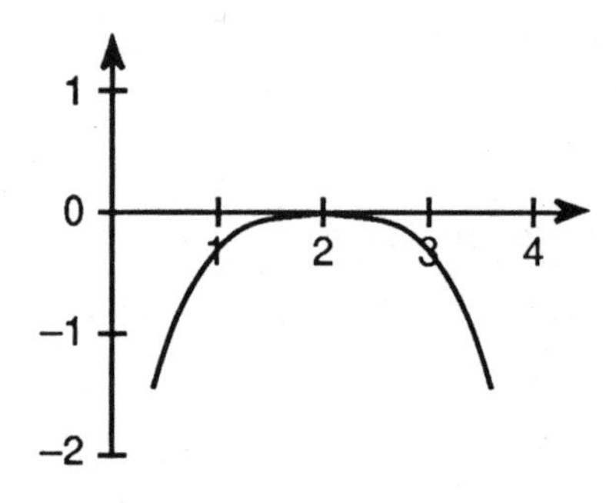

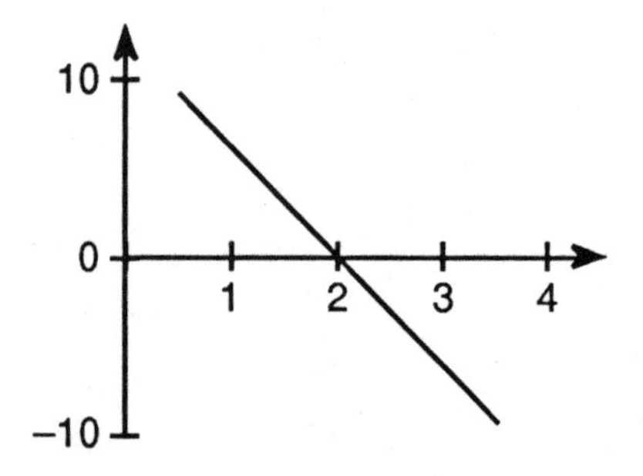

3.

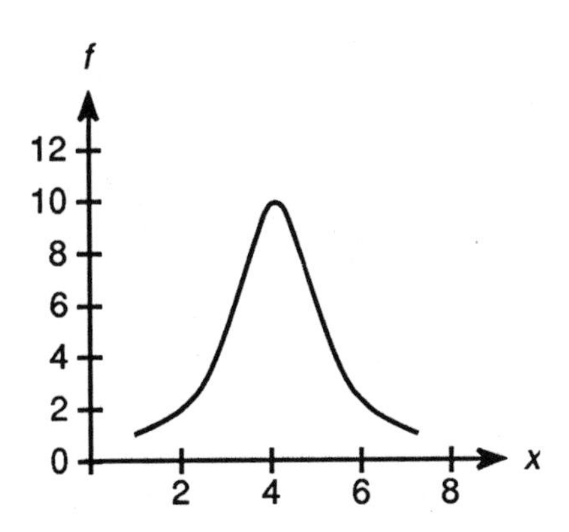

4.

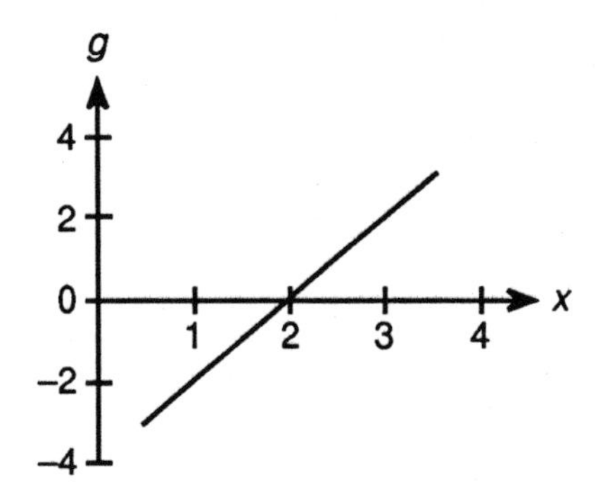

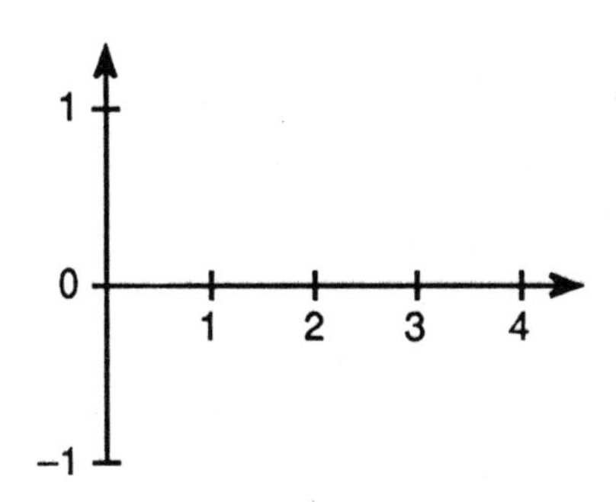

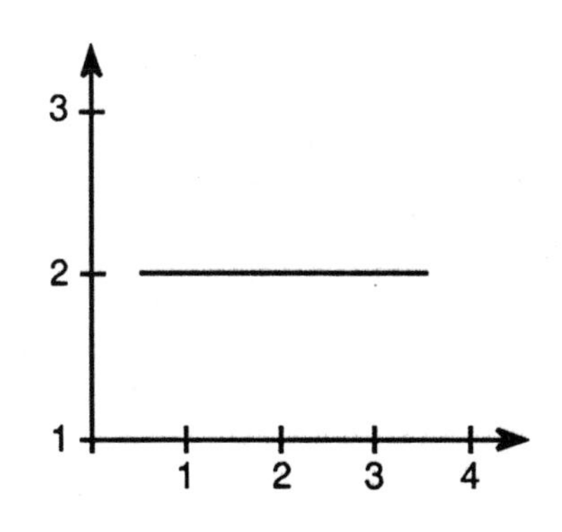

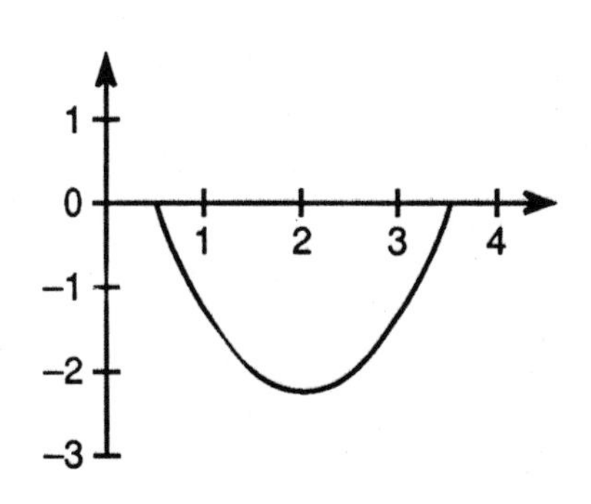

In exercises 5 and 6 the graphs of f and g are given at the top. The six graphs underneath are the graphs of f', f'', F, g', g'', G arranged randomly. Label each graph with the correct function.

5.

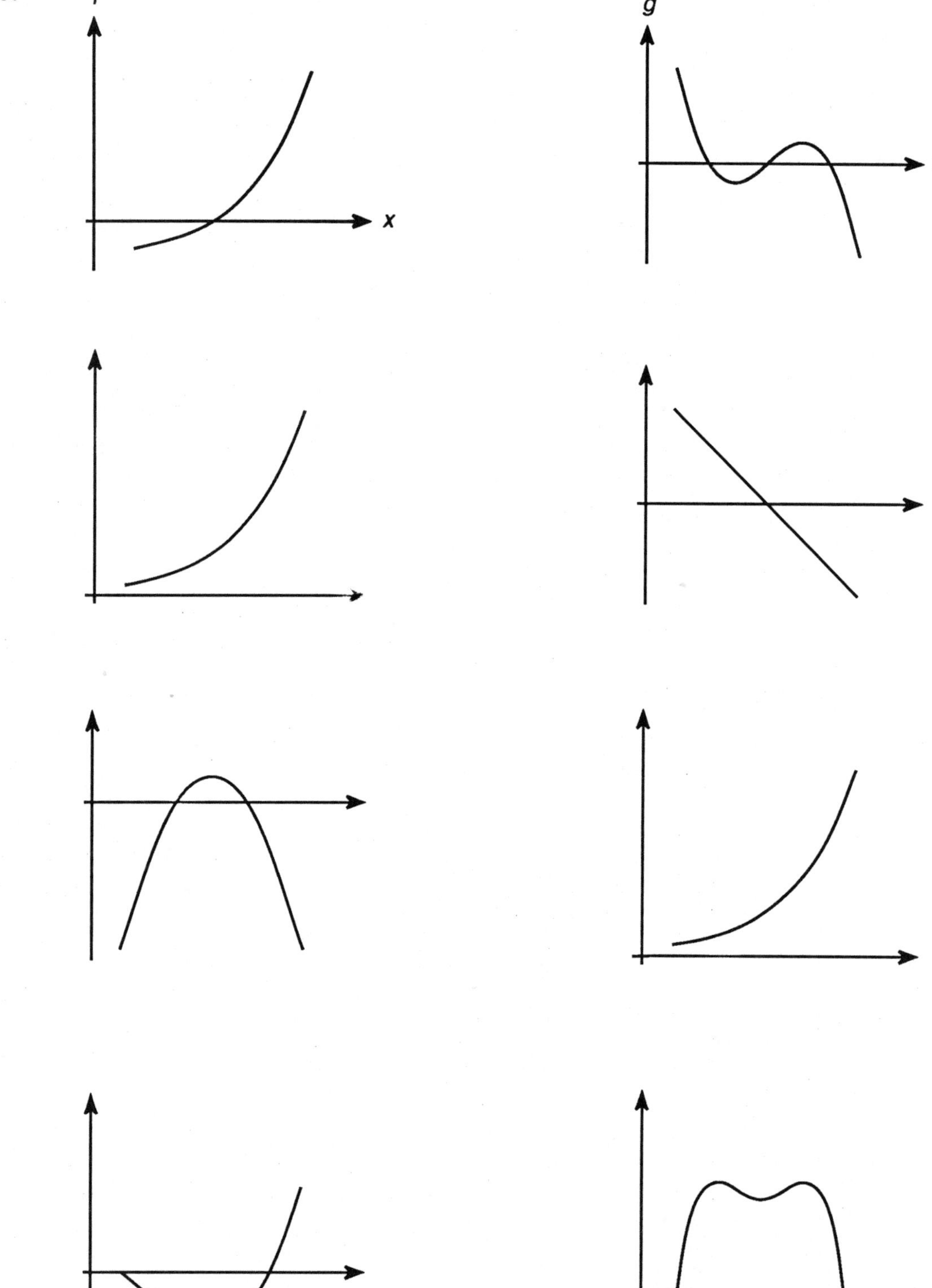

6.

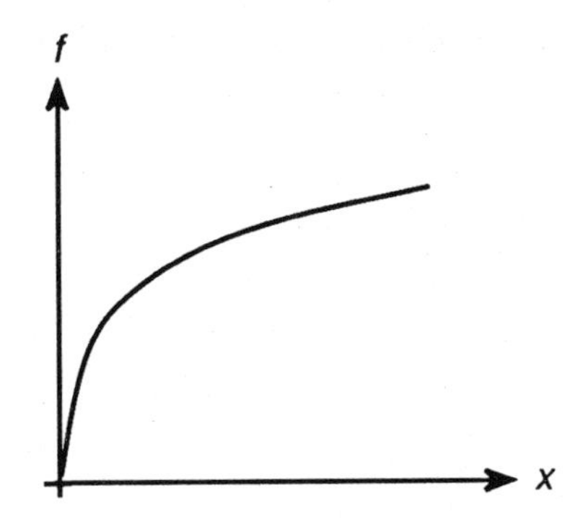

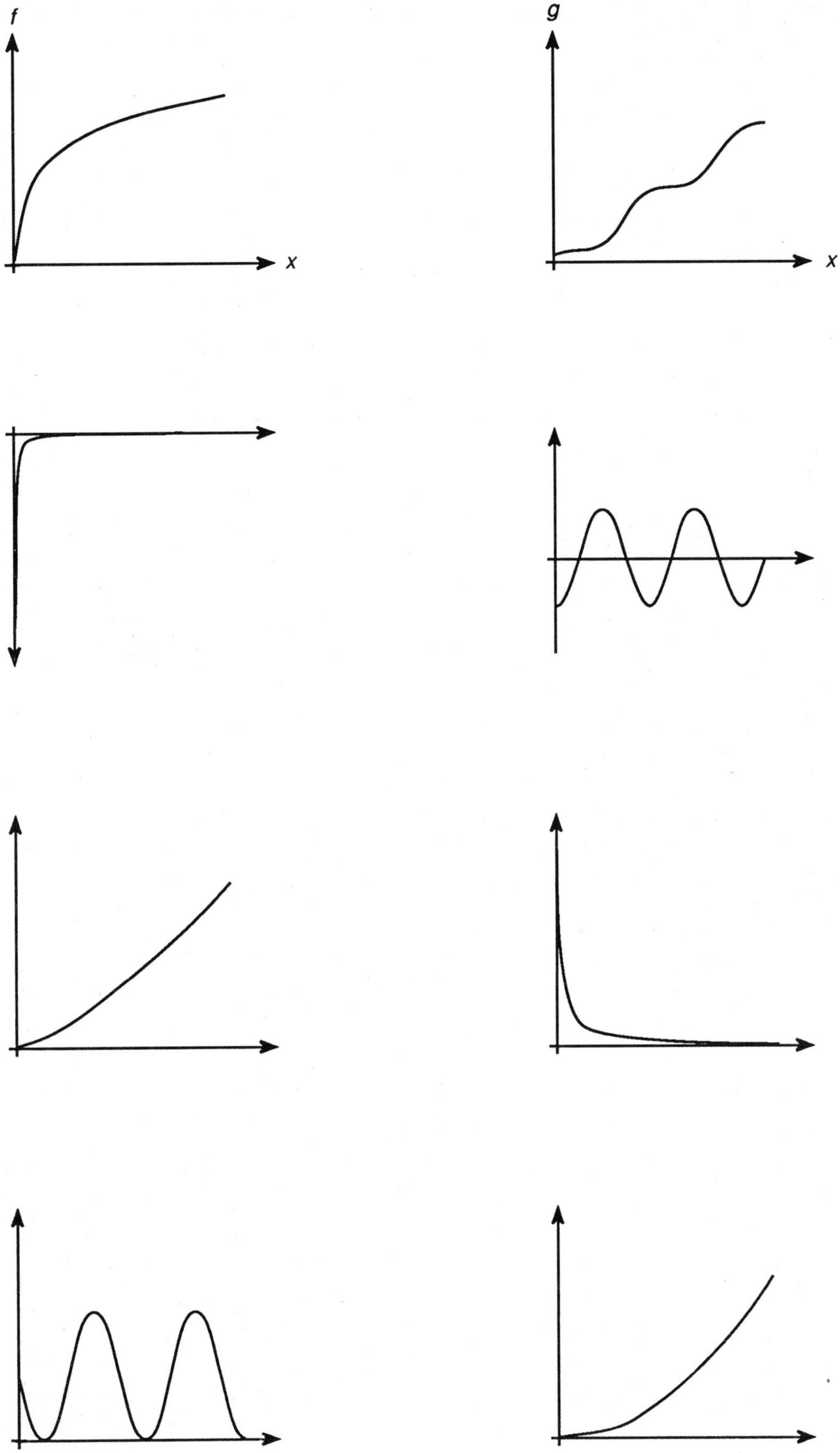

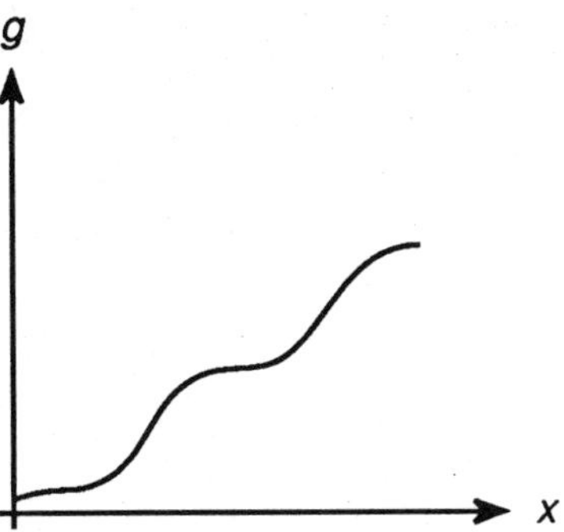

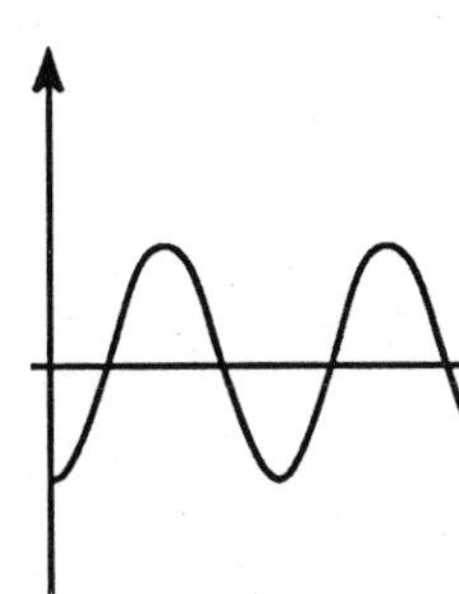

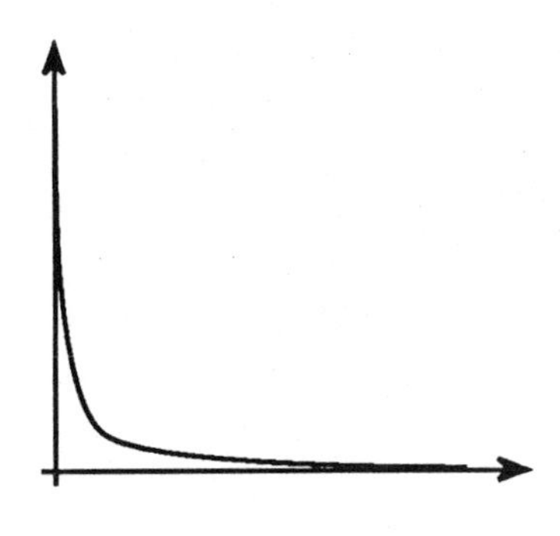

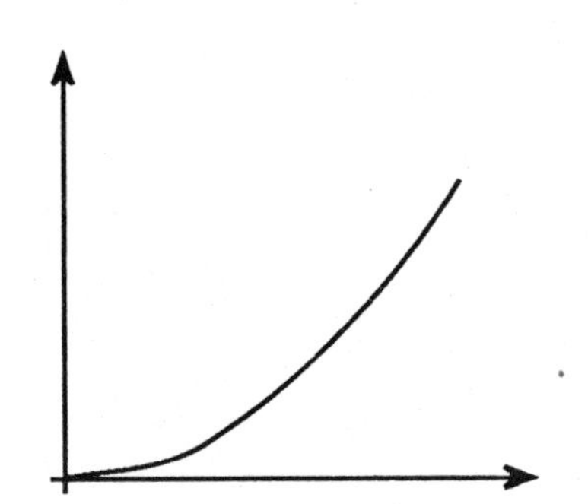

Answers to Exercises on Relating the Graphs of Functions and their Derivatives

Problem:	1.	2.	3.	4.	5.		6.	
	f	f	f	f	f	g	f	g
	f''	f'	F	f''	f''	g''	f''	g''
	f'	F	f''	f'	g'	f'	F	f'
	F	f''	f'	F	F	G	g'	G
					(f' and f'' can be reversed)			

Exercises on Newton's Method

1. Suppose we want to approximate the root of $f(x) = 0$ where the graph of f is shown to the right. We will use Newton's Method and require the approximation of the root to be in the neighborhood shown. Use a straightedge to graphically carry out Newton's Method and see how many steps it takes you to get inside the neighborhood.

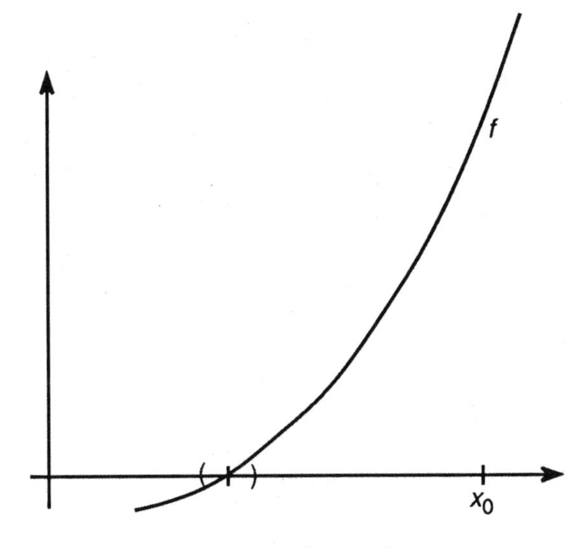

2. If the graph of f is concave upward, can you say anything about where the approximations will all be with respect to the rest? Does your answer depend on whether you start to the right or left of the root?

3. At the right is the graph of $f(x) = x + 1 + \sin x$. Is it easy or hard to solve $x + 1 + \sin x = 0$? What complication might arise in using Newton's Method if your first approximation was the x_0 shown?

4. The graph at the right shows another complication that can arise when using Newton's Method. Starting with the first approximation of x_0 for a root, follow the tangents shown and explain why a root will never be found using that x_0.

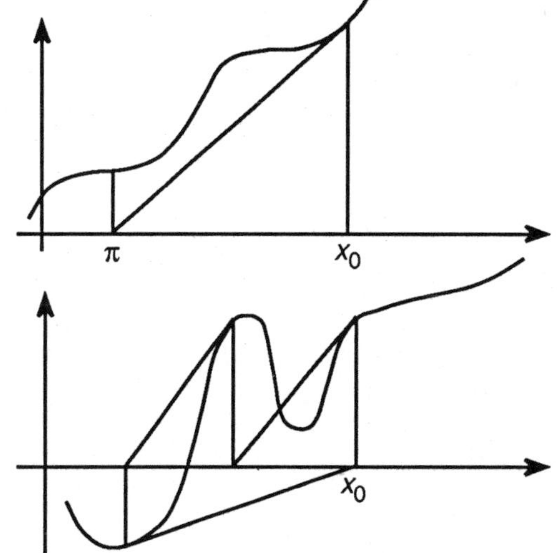

5. What conditions could be placed on f to insure the problems encountered in Exercises 3 and 4 wouldn't occur?

Remark. The examples in 3 and 4 are just a glimpse of unexpected behavior associated with Newton's Method. Computers with high-resolution graphics have revealed that if the method is applied to certain cubic polynomials of a complex variable $z = x + iy$, the situation is so wild that it is "chaotic." Chaos is an emerging science, with applications to a wide variety of problems in the physical, biological, and social sciences.

Answers to Exercises on Newton's Method

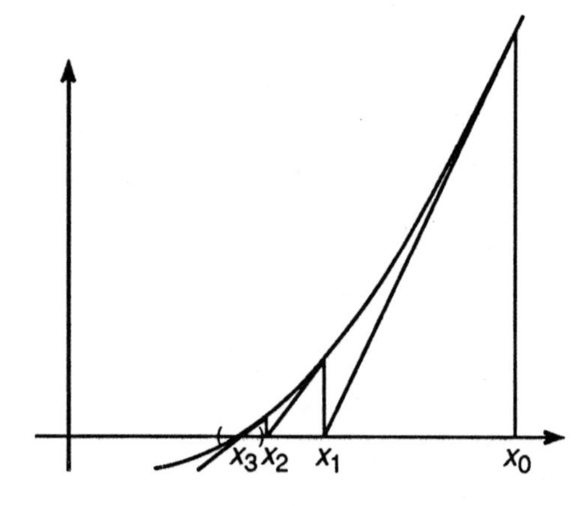

1. The figure shows that the third approximation will be inside the given neighborhood.

2. Each computed approximation will be too large.

3. No standard methods work in solving the transcendental equation $x + 1 + \sin x = 0$, so roots must be approximated. As shown, the second tangent line might be horizontal and not cross the x-axis anywhere. The condition $f'(x) \neq 0$ for any x would prevent that problem.

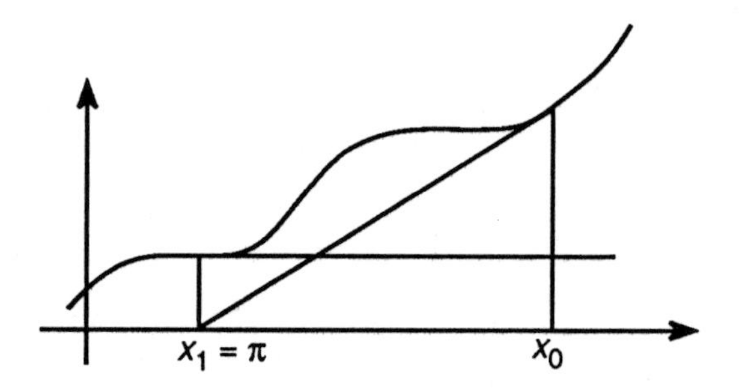

4. Approximations could conceivably cycle and not converge to a root. In our figure $x_3 = x_0$ and we are in a cycle.

5. Concavity would suffice; for example $f'' > 0$ (or $f'' < 0$) and $f' \neq 0$.

INVERSE FUNCTIONS

The Big Picture

Inverse Functions and their Derivatives

A good graphical understanding is a key to avoiding difficulties that students often have with inverse functions. If you understand these next pages, you will know almost all you need to know about inverse functions.

Start with a function whose graph has at most one point on each horizontal line, that is, with what we call a one-to-one function. Any strictly monotone function such as $x^3, e^x, \ln x$ or $ax + b, a \neq 0$ will have this property.

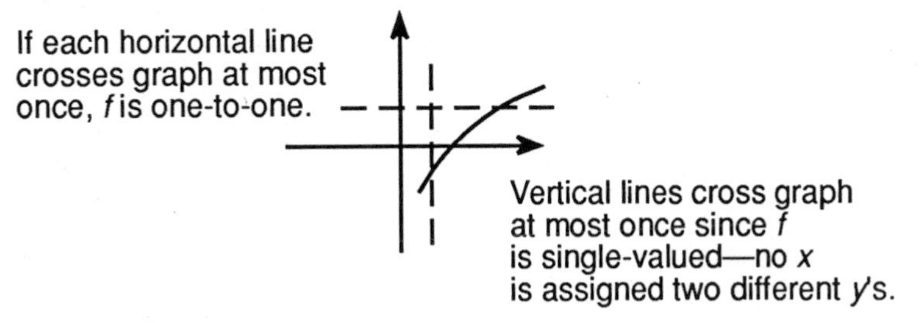

The graph of f^{-1} is obtained from the graph of f by reflection across $y = x$. The point (a, b) is on the graph of f if and only if (b, a) is on the graph of f^{-1}.

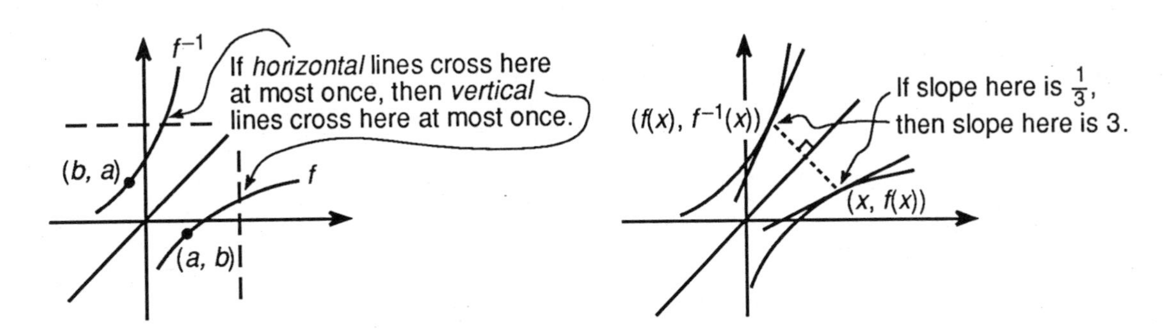

Since $(x, f(x))$ is on the graph of f, its reflection across $y = x$ is the point $(f(x), x)$. The point can also be written $(f(x), f^{-1}(f(x)))$ because x is the value f^{-1} assigns to $f(x)$. The slopes of the tangent lines to the graphs at these two points are reciprocals of each other. Keeping these simple graphical facts in mind will help you understand the mechanics of working with inverse functions.

Inverse functions simply reverse the assignments the original function makes. If f assigns b to a we write $f(a) = b$, and then f^{-1} assigns a to b and so we write $f^{-1}(b) = a$. Thus, (a, b) is on the graph of f if and only if (b, a) is on the graph of f^{-1}. It follows that $f^{-1}(f(x)) = x$ and $f(f^{-1}(x)) = x$.

Let's take a specific example: $f(x) = \sqrt[3]{x-1}$.

1. In order to have an inverse, the graph of f can only be crossed at most once by each horizontal line. Graph f to check that this is the case.

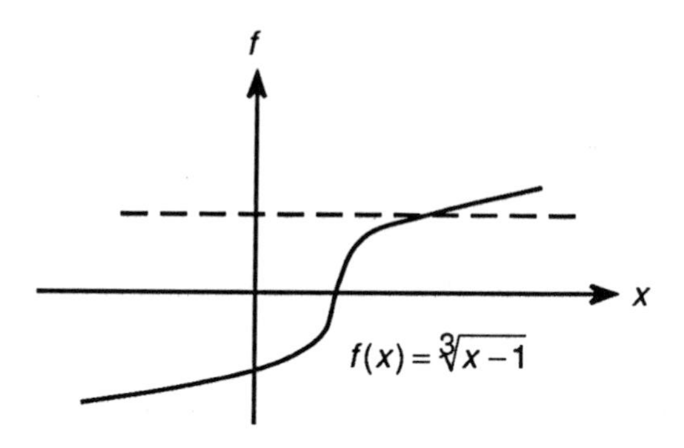

2. Reflect the graph of f across $y = x$ to get the graph of f^{-1}. To get a formula for f^{-1}:

   (a) Write $y = f(x)$. In our case $y = \sqrt[3]{x-1}$.

   (b) Switch x and y.
   $x = f(y)$, in our case $x = \sqrt[3]{y-1}$.
   (If (a, b) satisfies the first equation, then (b, a) satisfies the second.)

   (c) Solve for y to get $f^{-1}(x)$.

$$x = \sqrt[3]{y-1}$$

$$x^3 = y - 1$$

$$y = x^3 + 1 = f^{-1}(x)$$

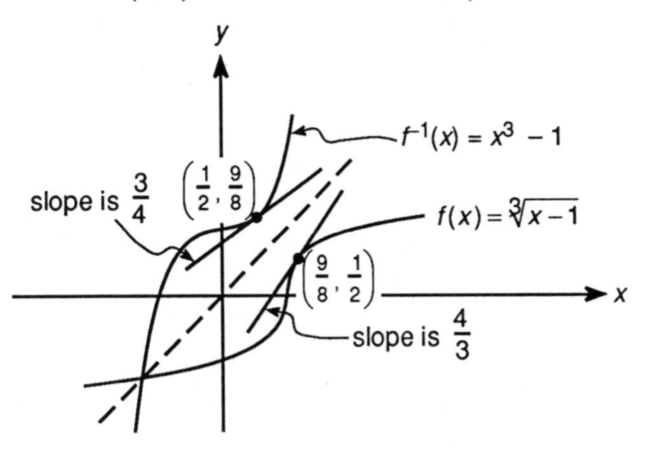

(Check: Does $f^{-1}f(x) = x$? $f^{-1}f(x) = (\sqrt[3]{x-1})^3 + 1 = x - 1 + 1 = x$. Yes!)

Notice the derivative of f is $f'(x) = \frac{1}{3}(x-1)^{-2/3}$. For example, when $x = 9/8, f(x) = \sqrt[3]{9/8 - 1} = 1/2$ and $f'(9/8) = \frac{1}{3}(\frac{1}{8})^{-2/3} = 4/3$.

Now note that $(f^{-1}(x))' = 3x^2$ and when $x = 1/2$ this gives a derivative of $3/4$. This illustrates the formula $(f^{-1})'(\frac{1}{2}) = 1/f'(9/8)$.

Let's now see how easy it is to justify some of this. *The graph of f^{-1} is the reflection of the graph of f across $y = x$.*

By definition (b, a) is on the graph of f^{-1} because (a, b) is on the graph of f. But (a, b) and (b, a) are reflections of each other across $y = x$. To see this, note:

   (a) The slope of the segment joining (a, b) and (b, a) is $\dfrac{a-b}{b-a} = -1$, so that segment is perpendicular to $y = x$.

   (b) The midpoint of the segment (a, b) and (b, a) is $\left(\dfrac{a+b}{2}, \dfrac{a+b}{2}\right)$, which lies on $y = x$.

It's that easy!

The slope of the tangent line to the graph of f at $(x, f(x))$ is the reciprocal of the slope of the tangent line to f^{-1} at the point $(f(x), x) = (f(x), f^{-1}(f(x)))$.

(a) Start with $f^{-1}(f(x)) = x$.

(b) Differentiate using the Chain Rule to get $f^{-1\prime}(f(x)) \cdot f'(x) = 1$ or $f^{-1\prime}(f(x)) = \frac{1}{f'(x)}$.
This last messy-looking equation says exactly the same thing the italicized statement does. This shows how the visual interpretation given by the italicized statement help us understand the derivative for the inverse function.

Now you know what you need to know about inverse functions. Before we end, here are two important observations:

1. f^{-1} has a derivative and is related by a simple formula to the derivative of f. This shows that nice properties of f will be "inherited" by its inverse function f^{-1}. Also note that if f is continuous, then so will f^{-1}; just think of their graphs. This observation is a cousin to statements like: "If f and g are both continuous, so is $f + g$," that is, nice properties of f and g will be inherited by their sum.

2. The notation "f^{-1}" is unfortunate, since it has nothing to do with reciprocals, for which we use the same notation when we write $2^{-1} = 1/2$. Don't fall into this common trap that was set for us years ago! If we used something like $\hat{f}$ for the inverse function, we would never confuse it with the reciprocal and we could write $\hat{f}'(f(x)) = \frac{1}{f'(x)}$. However, standardized notation must be observed even when we could improve it. Just be cautious and alert.

Remark: We have assumed f^{-1} is differentiable at the value $f(x)$, and in fact this is not too difficult to prove. The Chain Rule approach, while not a proof, has the advantage of quickly leading to the correct differentiation formula for f^{-1}. It should be geometrically obvious that f^{-1} is not differentiable at a point $b = f(a)$ for which $f'(a) = 0$.

The Big Picture

Implicit Differentiation

We know how to differentiate $y = f(x) = x^2 + \sin x^3$. We have rules to apply when we have y (or f) given *explicitly* in terms of x. *Implicit differentiation* applies in a different setting. If we have an expression like $xy + y \sin x = x^3$, we can solve for y to get $y = \dfrac{x^3}{x + \sin x}$ and use the usual explicit rules to get $\dfrac{dy}{dx}$. However, if we have something like $xy^2 + e^y \sin x = \pi$ we cannot solve for y. The expression has some sort of graph that may not even define a function, because for a given x more than one y works in the equation. In fact, for our example, we can see that the graph passes through $(\pi, 1)$ and $(\pi, -1)$ because $x = \pi$, $y = 1$, and $x = \pi, y = -1$ both satisfy the equation. Can we find the slope of the curve at $(\pi, -1)$ even if we can't write y as an explicit function of x? We haven't tried to differentiate a function we can't even write down! All we have encountered are well-defined, single-valued functions, but now for $x = \pi$ we have two ys determined.

We have some sort of curve given by $xy^2 + e^y \sin x = \pi$ that wanders through both $(\pi, 1)$ and $(\pi, -1)$.

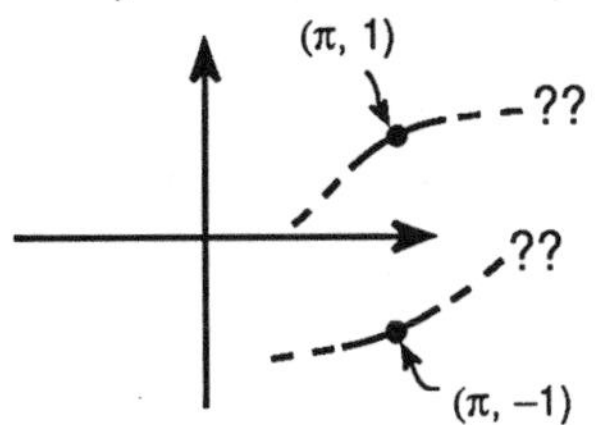

What is the slope of the tangent to the curve at these points? We proceed as follows: For xs near π there is some function y determined (y is the dependent variable, x is independent) if we restrict our y values to be close to $y = 1$. In other words, ignoring the part of the graph near $(\pi, -1)$, we have some single-valued function if we restrict our attention to the part of the graph near $(\pi, 1)$. We can't tell what that function is because we can't solve *explicitly* for y, but we can tell you what the slope of the tangent line is at $(\pi, 1)$ using *implicit differentiation.* If $xy^2 + e^y \sin x = \pi$, then $\dfrac{d}{dx}(xy^2 + e^y \sin x) = \dfrac{d}{dx}(\pi) = 0$ (since π is a constant). But what is $\dfrac{d}{dx}(xy^2 + e^y \sin x)$? Here we use the Chain Rule, *keeping in mind that y is some function of x.* Thus

$$\frac{d}{dx}(xy^2+e^y \sin x) = \frac{d}{dx}(xy^2)+\frac{d}{dx}(e^y \sin x) = (1\cdot y^2+x\frac{d}{dx}y^2)+(\frac{d}{dx}e^y)\sin x+e^y \cos x$$

$$= (y^2 + x(2y\frac{dy}{dx})) + (e^y\frac{dy}{dx}\sin x + e^y \cos x)$$

But all of this was equal to 0, remember? So we have

$$y^2 + 2xy\frac{dy}{dx} + e^y \sin x\frac{dy}{dx} + e^y \cos x = 0.$$

Solving for $\frac{dy}{dx}$ we get (just using algebra):

$$\frac{dy}{dx} = -\frac{(y^2 + e^y \cos x)}{2xy + e^y \sin x}.$$

Presto! We have found $\frac{dy}{dx}$ and don't even know what y is explicitly in terms of x. Notice the price we have paid: our expression for $\frac{dy}{dx}$ depends not only on x (as it always did for explicit functions), but on y also.

Now we can answer our original questions: What are the slopes of the lines tangent to the curve at $(\pi, 1)$ and $(\pi, -1)$? We simply plug in the values.

$$\left.\frac{dy}{dx}\right|_{(\pi,1)} = -\frac{(1 + e(-1))}{2\pi} = \frac{e-1}{2\pi} \approx 0.27$$

$$\left.\frac{dy}{dx}\right|_{(\pi,-1)} = -\frac{1 + e^{-1}(-1)}{-2\pi} = \frac{1 - e^{-1}}{2\pi} \approx 0.10$$

Thus we now know the slopes at which the curve passes through the two points $(\pi, 1)$ and $(\pi, -1)$.

The Big Picture

Maximum–Minimum

There are many settings in which we want to maximize or minimize some quantity. We like to minimize costs and maximize profits. In some situations we may want to minimize time spent doing something; in others we may want to maximize time. We may want to maximize or minimize volumes of figures given to us in some way. How to accomplish this fits into a general process called *optimization*, that is, how to make optimal decisions. The derivative gives us one important tool in the optimization tool kit.

In upcoming applications, you will be given (although you may have to work to get it) a function f that expresses costs, profits, volumes, or other quantities you wish to maximize or minimize. The process of optimizing with calculus tools is easily visualized–it is simply a matter of finding the peaks and valleys on the graph of f.

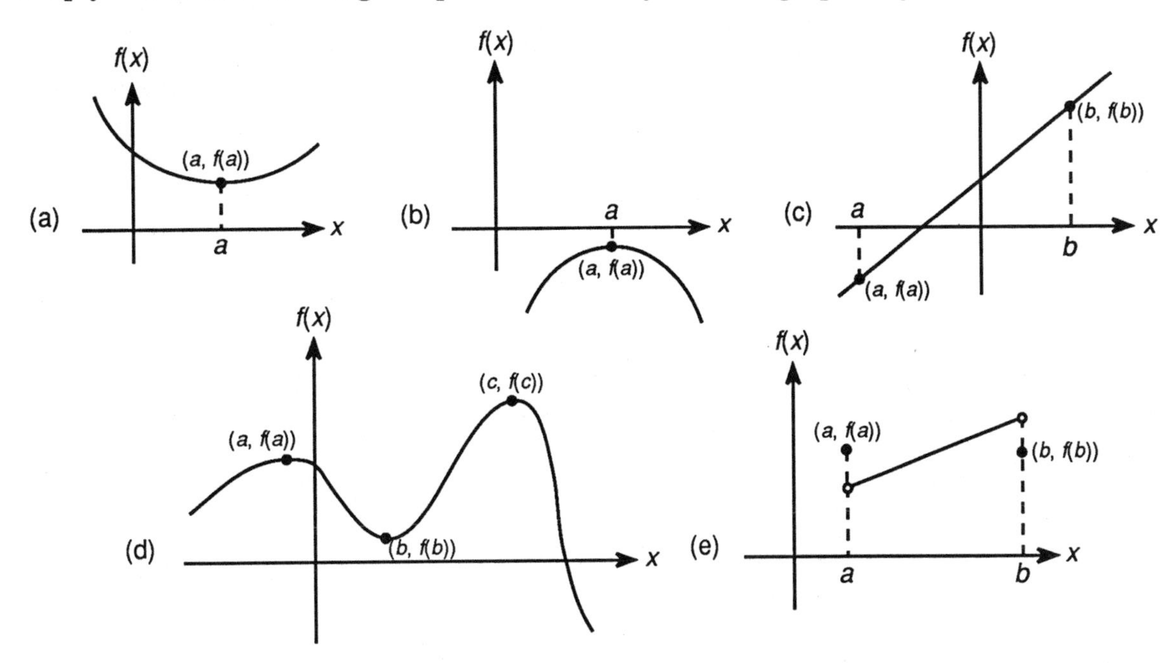

Consider the graphs shown. In graph (a), f has a minimum at $x = a$ and that minimum value is $f(a)$. By the way, what kinds of functions do you know that have graphs like this? Many students are sloppy in not carefully making a distinction between the value a of the independent variable, where the minimum occurs, and $f(a)$, which is the minimum value. Teachers are also sloppy in not correcting students incorrect usage of terminology, partly because whether they are careful to say it or not, it is always visually obvious what they mean. So be careful to say what you mean when you write up your solutions.

For the function f whose graph is shown in (b), f has maximum value at $x = a$. (This is negative, but that is as big as f gets anywhere.) In (c) the function has no maximum or minimum value if we consider f over the domain of all real numbers. Wherever we are on the graph, there are always bigger or smaller values. However, in many real situations we only want to consider a function over a domain of some closed interval $[a, b]$. So in (c), if we are asked to maximize f on $[a, b]$, the maximum value is $f(b)$ while the minimum value is $f(a)$. Notice also that f has no maximum or minimum on the *open* interval domain (a, b). (Why not?)

The Maximum Value Theorem says that continuous functions always take a maximum value on a closed interval domain $[a, b]$. The same is true, of course, for minimum value. Notice that that statement is not true for all continuous functions on *open* intervals (a, b). Also, non-continuous functions don't have to have maximum or minimum values on closed intervals $[a, b]$, as the graph in (e) shows.

The graph in (d) presents an interesting situation. The value $f(a)$ is not an ***absolute*** maximum because $f(c)$ is larger. But $f(a)$ is larger than any other values $f(x)$ for x near a. In this case we say f takes a ***relative*** or ***local*** maximum value $f(a)$ at $x = a$. The value $f(c)$ is an ***absolute*** or ***global*** maximum. Similarly, we say $f(b)$ is a local minimum value, but it is not a global minimum value because f takes smaller values elsewhere.

Suppose we have a "nice" function f on a closed interval $[a, b]$; that is, f is a function continuous on $[a, b]$ with a derivative at all points in (a, b).

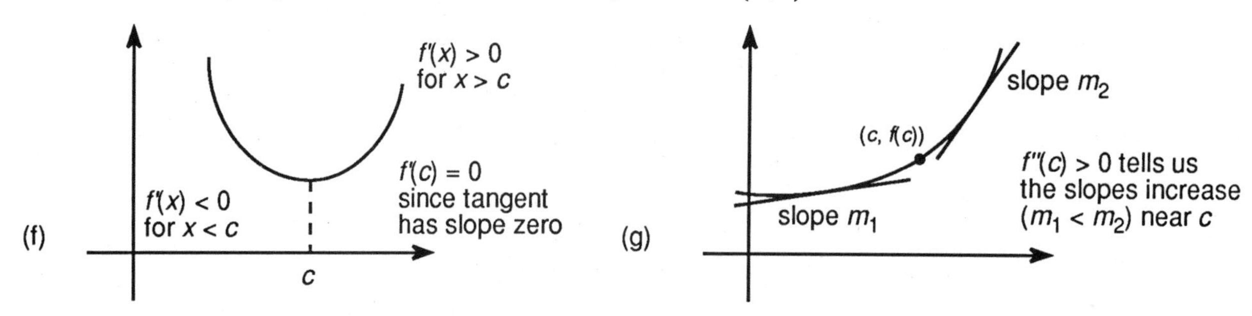

If we are asked to find its maximum and minimum values on $[a, b]$, how do we proceed? If the maximum or minimum occurs at an interior point in (a, b), it must occur where the tangent line to the graph is horizontal, that is, at points where f' is zero. So we need to do two things.

1. Find *all* $c \in (a, b)$ where $f'(c) = 0$ (by setting $f'(x) = 0$ and solving for x) and evaluate $f(c)$.

2. Compute $f(a)$ and $f(b)$, the values at the endpoints. Then we simply sort through the values $f(a), f(b)$ and $f(c)$ (for all such cs) and pick the largest and smallest values. Always keep a visual picture in your mind of what you are doing .

The functions f' and f'' are both useful in sorting out maximum and minimum values. If $f'(x) < 0$ for $x < c, f'(c) = 0$, and $f'(x) > 0$ for $x > c$, then $f(c)$ is a minimum value. (This is the "first derivative test.") Nothing more needs to be said; see the graph in (f). The use of f'' is a little more subtle, but it also tells us about the graph of f near $(c, f(c))$ if $f''(c) \neq 0$. Here is an important point to understand: *f'' tells us the same things about f' that f' tells us about f itself.* Where $f' > 0, f$ is increasing; that is, slopes of tangent lines are positive. Assuming f'' is continuous and $f''(c) > 0$, we see that $f'(x)$ is increasing for x near c. This means that the graph of f bends upward, or is *concave upward* near c. Similarly, where $f'' < 0$ the graph of f is *concave downward.* Let's summarize:

$f(x) > 0$	graph of f is above the horizontal axis at x
$f(x) = 0$	graph of f lies on the horizontal axis at x
$f'(x) > 0$	f is increasing at x, graph of f runs southwest to northeast
$f'(x) < 0$	f is decreasing at x, graph of f runs northwest to southeast
$f'(x) = 0$	graph of f has a horizontal tangent at $(x, f(x))$
$f''(x) > 0$	graph of f is concave upward
$f''(x) < 0$	graph of f is concave downward
$f''(x) = 0$	further tests must be run

Thus, if $f'(c) = 0$ and $f''(c) < 0$, then $f(c)$ must be a local or global maximum value (the graph has a horizontal tangent and is concave downward; this is the "second derivative test"). If you have all of this information *visualized*, you will find the process of finding maximum and minimum values very easy.

See how you do on these questions about a function f whose graph is shown.

Probes

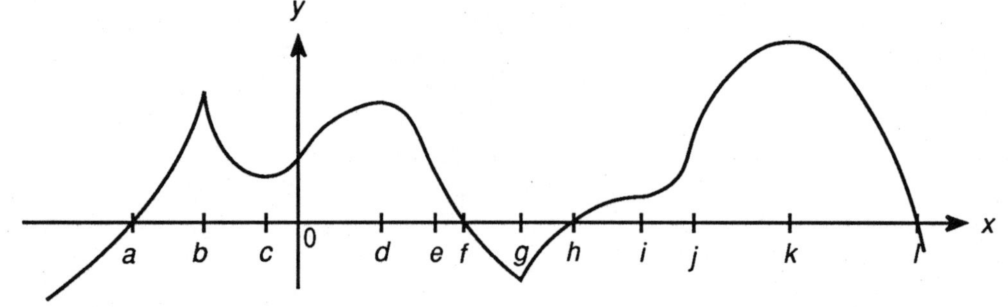

Maximum–Minimum

1. Are there any points of discontinuity?
2. At what values x is $f(x) = 0$?
3. At what values x does $f'(x)$ not exist?
4. At what values x is $f'(x) = 0$?
5. Is $f'(e)$ positive or negative?
6. At what values x are you sure $f''(x) = 0$?
7. If $f''(h)$ exists and is not zero is it positive or negative?
8. At what value x is $f(x)$ a local or global maximum value, even though $f'' > 0$ for all points of the domain near but not equal to x?
9. What are local maximum values that are not global in $[a, j]$?
10. What is the global maximum value over the domain of the whole real line?
11. For what value x is $f'(x) = 0$, yet $f(x)$ is neither a relative maximum nor minimum value?

Here are three examples to keep in mind to see why the second derivative test fails when $f''(x) = 0$.

(a) $f(x) = x^4$
$f'(0) = f''(0) = 0$
f has a minimum at $x = 0$.

(b) $f(x) = -x^4$
$f'(0) = f''(0) = 0$
f has a maximum at $x = 0$.

(c) $f(x) = x^3$
$f'(0) = f''(0) = 0$
f has neither maximum nor minimum at $x = 0$.

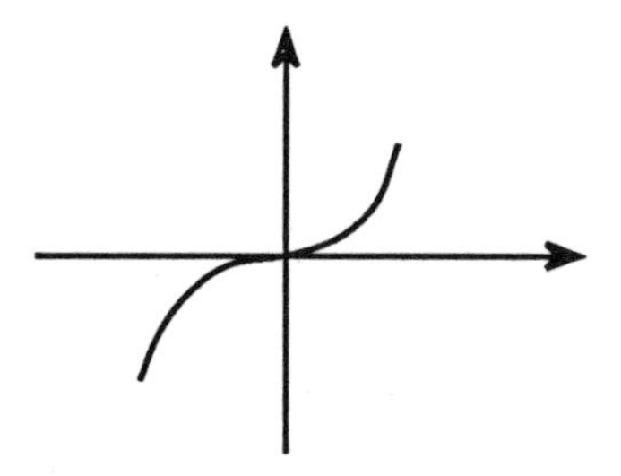

One final comment about optimization: For a linear function, whose graph is a line, the maximum or minimum occurs at the endpoints (or boundary) of $[a, b]$, and the derivative is of no help. In three dimensions, the maximum or minimum z value on a plane does not exist unless we restrict ourselves to some subset of the plane. For example, c is the maximum z value for the portion of the plane that lies above the shaded portion of the x, y plane shown. Any such maximum value will have to lie above or below a point that is on the *boundary* of the portion of the x, y plane to which you restrict your attention. Finding such maximum and minimum values in these (and even higher dimensional) situations is another important and well-developed area of optimization. Such "linear" optimization problems are solved with non-calculus methods. Sophisticated computer software has been developed for such problems. Full college courses can be taken to learn how this optimization is done. Such studies are a part of the field called *operations research* that blossomed with the extensive use of computers during World War II.

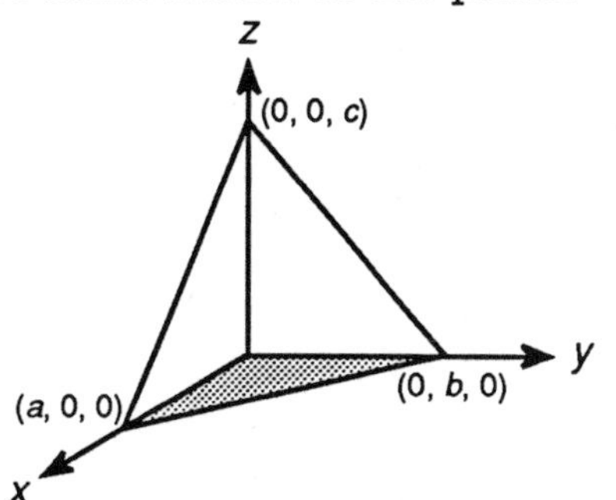

In multivariable calculus courses you will also learn how to find maximum and minimum values on curved, non-planar surfaces.

In summary, your problems on maximum and minimum values are the first examples of optimization. The whole story is quite a long one, using vastly different tools depending on the situation.

Answers to Maximum-Minimum Probes

1. no
2. a, f, h, l
3. b, g
4. c, d, i, k
5. negative
6. 0, e, i, j
7. negative
8. b
9. f(b), f(d)
10. f(k)
11. i

Constrained Max-Min Problems Solved by Graphical Analysis

The Picture-Viewing Problem

In 1471, the mathematician Regiomontanus posed the following problem:

At what distance from a wall does a picture appear tallest?

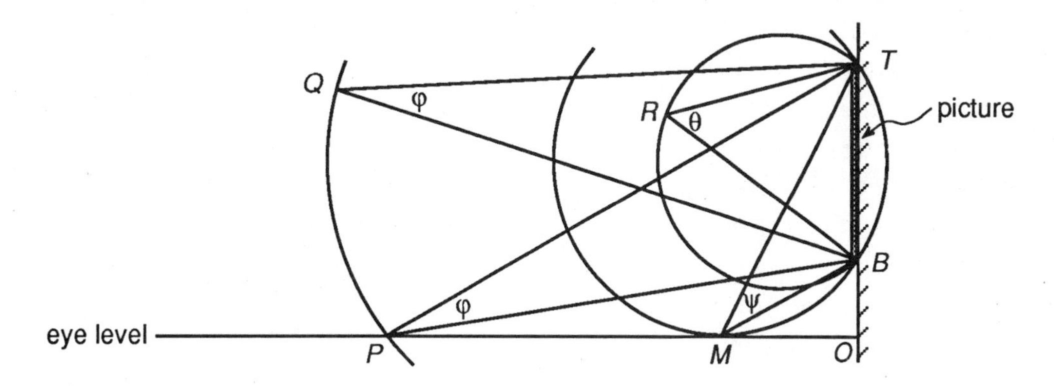

This problem is one of the earliest *extreme* problems encountered in the history of mathematics.

Let's consider a solution that does not use calculus directly, but does show the value of using tangents to various curves.

As the figure shows, our eye level is below the bottom B of the picture. At point P we see the picture with angle $\varphi = \angle TPB$. Indeed, from geometry we would see the picture with the same angle φ from any point, such as Q, on the circular arc passing through P, B, and T. With a ladder, one could see the picture from point R with angle θ, and certainly $\theta > \varphi$. The viewing angle gets larger as the circles through B and T get smaller. Without a ladder, the largest viewing angle is at M, the point where a circle through B and T is tangent to eye level. By the intersecting secant lines theorem* we have $(OM)^2 = OB \cdot OT$. Thus, we should stand at a distance $OM = \sqrt{OB \cdot OT}$ from the wall. Don't leave home without your ruler and calculator–at least if you are headed to the art museum.

*If two secant segments are drawn from an external point, the product of the lengths of one secant segment and its external part is equal to the product of the lengths of the other secant segment and its external part.

*Another Max-Min Problem**

Suppose a function f is decreasing and its graph in the first quadrant is concave downward between $(0, f(0))$ and $(x, 0)$. Choose any point P on the graph and draw the tangent to f at P. Label its x and y-intercepts A and B. Consider rectangle $OCPD$ and triangle ABO. *With no other information about f* we can show that the point P that maximizes the area of $OCPD$ also minimizes the area of ABO and bisects segment $\overline{AB}$! (In fact, a point P that has any one of the three properties must have the other two.)

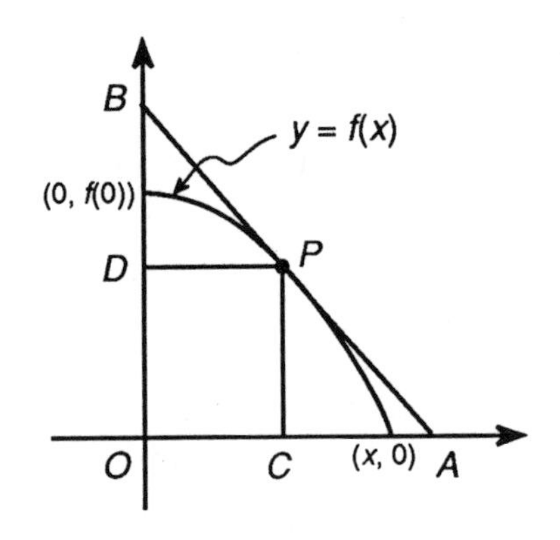

There is a beautiful graphical proof of this result whose actual computation is minimal. We will need the following fact, a routine calculus exercise that is included in the following exercise set.

(∗) Let ℓ be any line tangent to the graph of $y = \frac{k}{x}$ in the first quadrant. Let A and B be the two intercepts of ℓ and P be the point of tangency. Then P bisects AB.

Now notice that all points on the graph of $y = \frac{k}{x}$ determine rectangles and triangles of constant area. If (x_0, y_0) is any point on the graph of $y = \frac{k}{x}$, then $x_0 y_0 = k$, and so the area of $OCPD$ is k. Using (∗), the area of the triangle ABO is $\frac{1}{2}(2x_0)(2y_0) = 2x_0 y_0 = 2k$. Thus, the areas do not change as P moves on the graph of $y = \frac{k}{x}$.

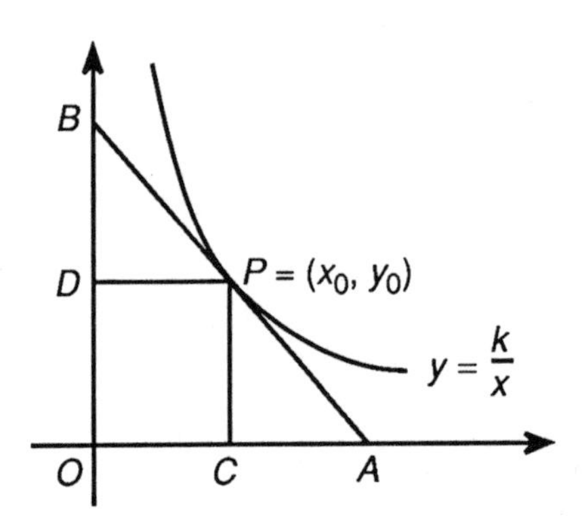

*See Herbert Bailey, "A Surprising Max-Min Result," *College Math Journal 18* (1987), 225-229.

Next, think of having the family of hyperbolas given by $y = \frac{k}{x}$ drawn in the first quadrant for various k. If we want a point P on the graph of f that maximizes the area of the corresponding rectangle $OCPD$, it should come from the curve of the family $y = \frac{k}{x}$ with largest possible k, say k_0. As we see in the figure, that point will be where the graph of f and the hyperbola have a common tangent. Thus, by $(*)$, if P maximizes the area of $OCPD$, then it also bisects $\overline{AB}$.

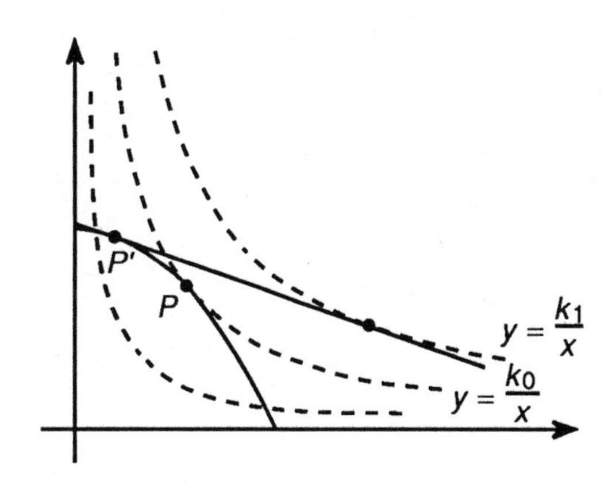

Finally, note that for any other point P' on the graph of f the corresponding tangent line to f at P' is also tangent to a hyperbola $y = \frac{k_1}{x}$ with $k_1 > k_0$. Here we use the concave downward assumption. Recall: P' determines a triangle of area $2k_1$, while P gives one of area $2k_0$, and therefore P has also given the triangle of minimal area.

Exercises on Constrained Max-Min Problems Solved by Graphical Analysis

1. Show that if l is tangent to the graph of $y = \frac{k}{x}$ at P, then P bisects $\overline{AB}$.

2. An x by y rectangle has perimeter $P = 2x + 2y$ and area $A = xy$. Thus the points along the lines $x + y = c, c > 0$, correspond to rectangles of constant perimeter $2c$, and points along any hyperbola $xy = k$ correspond to rectangles of constant area

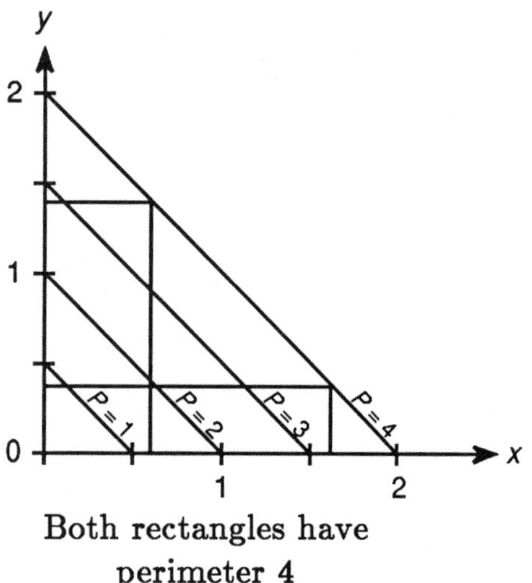

Both rectangles have perimeter 4

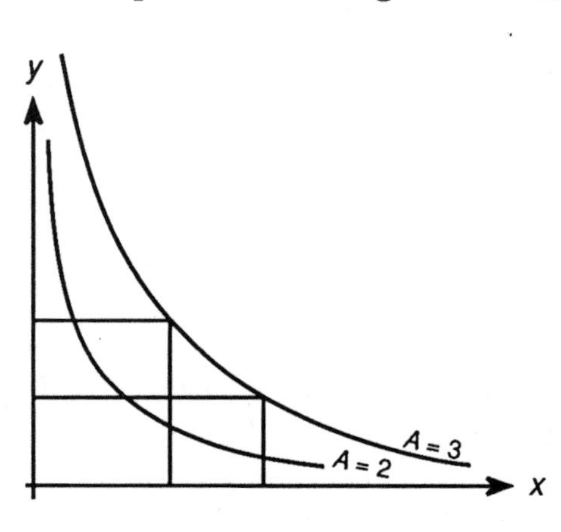

Both rectangles have area 3.

Relate the following problems to the curve systems shown above. It will be useful to imagine both systems on the same graph.

(a) Among all rectangles of area $A = 4$, which has the least perimeter?

(b) Among all rectangles of perimeter 12, which has the most area?

3. Use a graphical analysis similar to 2(b) to analyze the problem.

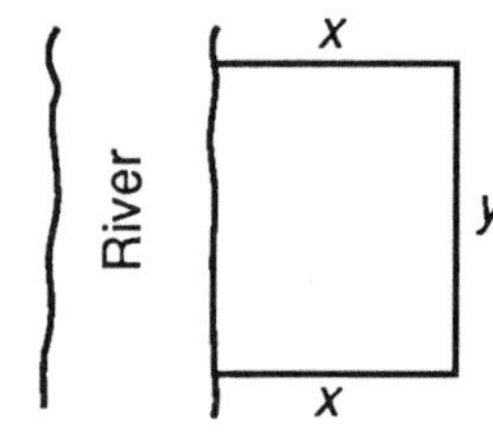

Given 4 kilometers of fencing, what are the dimensions of the x by y rectangular plot along a straight river bank that encloses the most area?

4. Carry out the actual computations below to illustrate our general result.

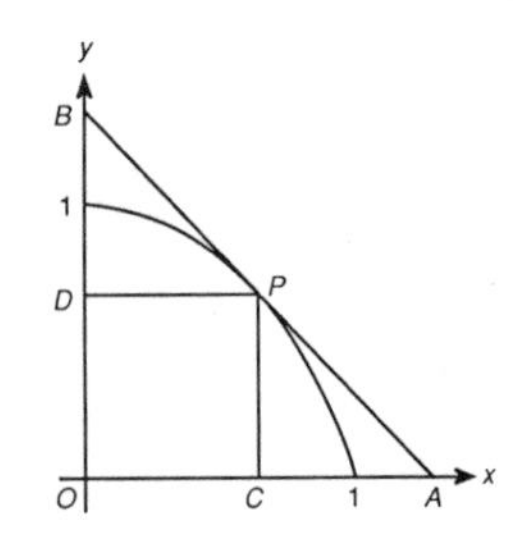

Show that the point P on the graph of $f(x) = 1 - x^2, 0 \leq x \leq 1$ that maximizes the rectangular area $OCPD$ also gives the minimum area of the triangle OAB formed by the tangent line to $y = 1 - x^2$ at P. Also show P is the midpoint of $\overline{AB}$.

Answers to Exercises on Constrained Max-Min Problems

1. Since $y' = -\frac{k}{x^2}$, the slope of the tangent line at P is $\frac{-k}{x_0^2}$. The equation of l is $y - y_0 = -\frac{k}{x_0^2}(x - x_0)$, which has intercepts $(0, 2y_0)$ and $(2x_0, 0)$. Therefore P bisects $\overline{AB}$.

2. (a)

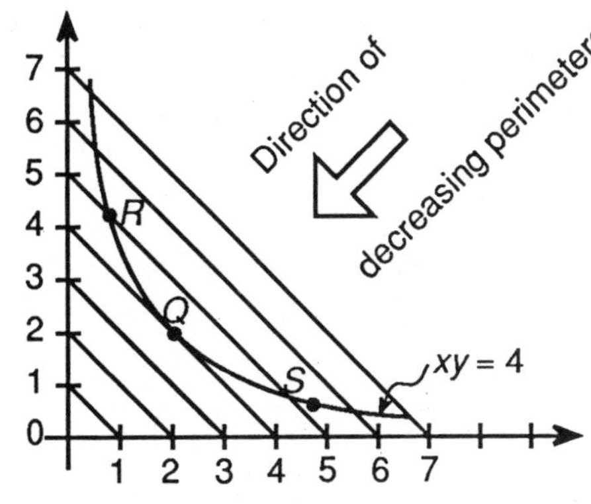

Let $Q = (x_0, y_0)$ be the unique point on $xy = 4$ that is tangent to a line in the family $x + y = c$. Since moving away from Q(say to R or S) increases the perimeter, Q corresponds to the minimum point. The lines $x + y = c$ have slope -1, so on $y = 4/x$ we want $y_0' = -4/x_0^2 = -1$. Thus $x_0 = 2$, and so $y_0 = 4/x_0 = 4/2 = 2$. The minimum perimeter is therefore $P = 2x_0 + 2y_0 = 8$.

(b)

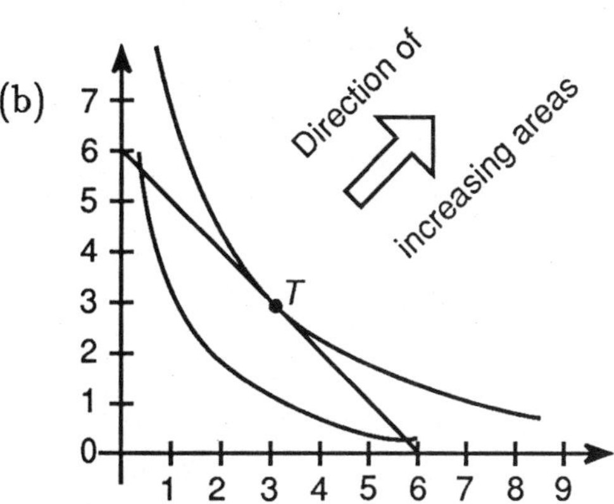

The point $T = (x_1, y_1)$ on $x + y = 6$ of maximum area corresponds to the unique point that is tangent to the hyperbola $y = k/x$. Thus $y_1' = -\frac{k}{x_1^2} = -1$. Since T is on $y = k/x$, we also have $y_1 = k/x_1$. Thus $x_1^2 = k = x_1 y_1$ and so $x_1 = y_1$. Since T is also on $x + y = 6$, we get $x_1 + y_1 = 2x_1 = 6$. Therefore, $x_1 = 3 = y_1$.

3.

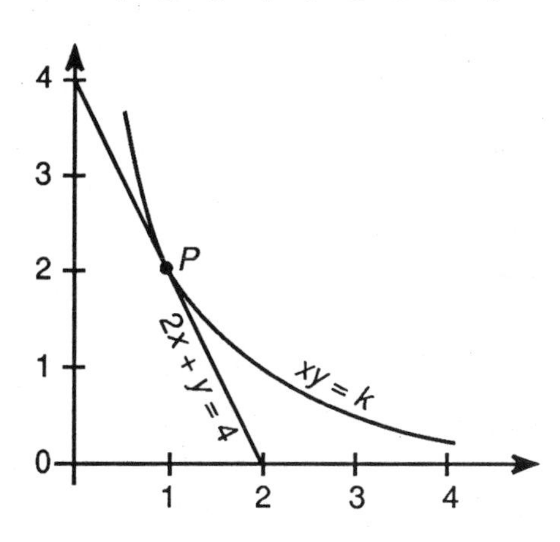

The point $P = (x_0, y_0)$ on $2x + y = 4$ of largest area occurs where the slope of $2x + y = 4$, namely -2, matches that of the hyperbola, $y_0' = -\dfrac{k}{x_0^2} = -\dfrac{k}{x_0}\dfrac{1}{x_0} = -y_0/x_0 = -2$. Thus, $y_0 = 2x_0, 2x_0 + y_0 = 4, 2x_0 + 2x_0 = 4$, so $x_0 = 1$, $y_0 = 2$. The max area is $x_0 y_0 = 2$ square kilometers.

4.

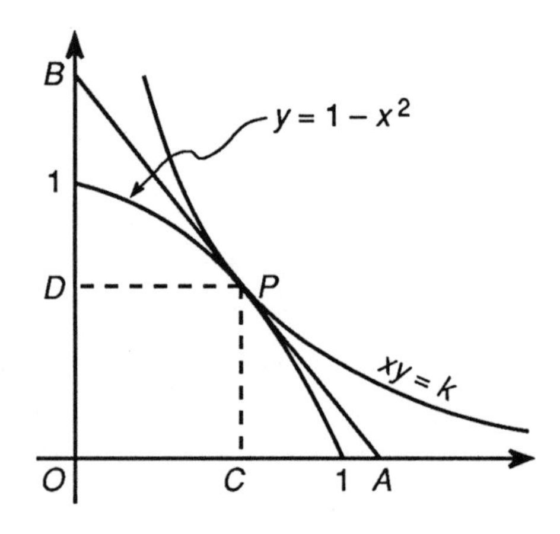

The area of $OCPD$ is largest at the unique point $P = (x_0, y_0)$ where the slope of $y = 1 - x^2$ matches that of the hyperbola $y = k/x$. Thus $f'(x_0) = -2x_0 = -\frac{k}{x_0^2} = -\frac{y_0}{x_0} = -\frac{(1-x_0^2)}{x_0}$. Solving for x_0, we find $x_0 = 1/\sqrt{3}$, and hence $y_0 = 1 - x_0^2 = 2/3$. Since $CA = y_0/(-f'(x_0)) = (k/x_0)/(k/x_0^2) = x_0 = CO$, it follows that $PA = PB$. That is, P is the midpoint of AB, and let the hyperbola $y = k_1/x$ be tangent to EF at $R = (x_1, y_1)$. If $S = (x_1, 0)$, then $SE = SR/(-y_1') = y_1/(k_1/x_1^2) = x_1 = SO$. Since S bisects OE, R bisects EF, and so area $OEF = \frac{1}{2}(2x_1)(2y_1) = 2x_1y_1$.

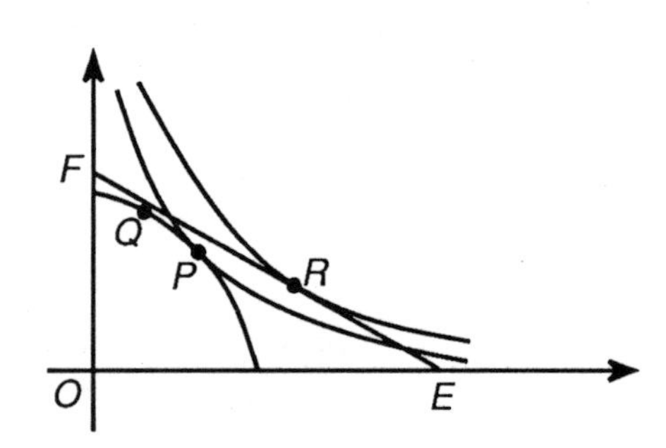

Let EF be tangent to $y = 1 - x^2$ at Q, where $Q \neq P$, but $x_1y_1 > x_0y_0$, so area $OEF > 2x_0y_0 =$ area OSB.

Periodic Billiard Paths

An Application of the Maximum Value Theorem

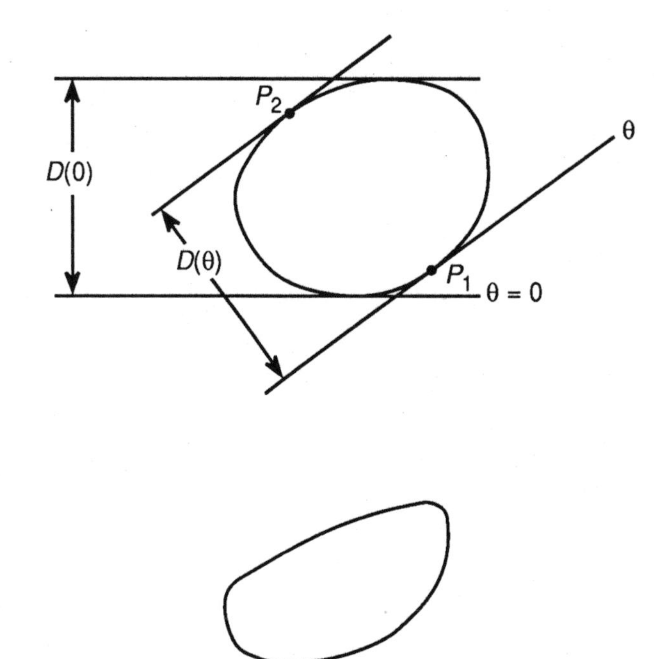

A closed convex curve C is "smooth" if it has a continuously-turning tangent (no sharp corners). For a direction $\theta, 0 \leq \theta \leq \pi$, the width of the curve in direction θ is the distance between the two parallel lines tangent to C and running in direction θ. When C is a smooth, closed, convex curve, $D(\theta)$ is continuous for $\theta \in [0, \pi]$, so the Maximum Value Theorem for continuous functions on closed intervals ensures that $D(\theta_0)$ is maximum for an appropriate choice of θ_0.

Exercises on Periodic Billiard Paths

1. Sketch in the parallel lines for the θ_0 that maximizes $D(\theta)$ for the curve above, and label the two points of tangency P_1 and P_2 as above.

2. Prove that the two tangent lines are perpendicular to the segment $\overline{P_1 P_2}$.

3. What would happen if a billiard ball was shot from P_1 toward P_2 and allowed to roll indefinitely?

The extension of the Maximum Value Theorem can be used to show that among all the triangles with three vertices on C, there is at least one with maximum perimeter. The following argument, due to George Birkhoff, shows that such a triangle of maximum length is a "periodic billiard path." That means that a ball shot from one vertex toward another will bounce off the curve C so as to repeatedly trace out the triangle.

Let A, B, C be the vertices of a triangle inscribed in C that has maximum perimeter. We wish to show that a ball shot from point B to point C will be reflected to point A. Think of the family of ellipses with foci at A and B. One of that family passes through

C. If it does not have the same tangent as the curve $\mathcal{C}$ at point C, there is some larger member of the family of ellipses with a point on $\mathcal{C}$, say at C'. Now since $d_1' + d_2' > d_1 + d_2$, the triangle ABC' has greater perimeter than does triangle ABC, contradicting triangle ABC having maximal perimeter. Thus the ellipse with foci A and B passing through C is tangent to the curve $\mathcal{C}$ there, and the bounce from B to C will be directed to A. The same argument can be repeated at any vertex.

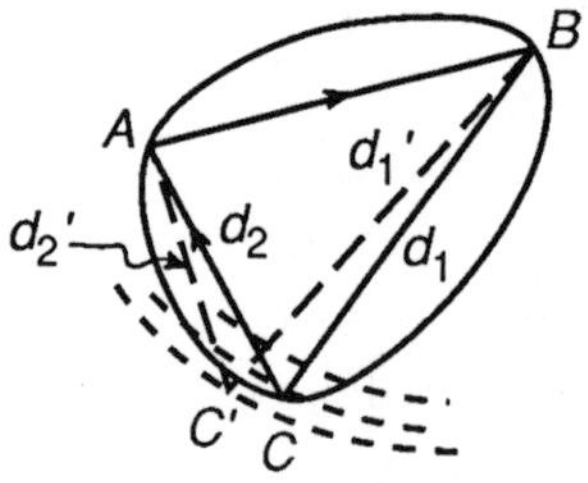

Answers to Periodic Billiard Paths Exercises

1.

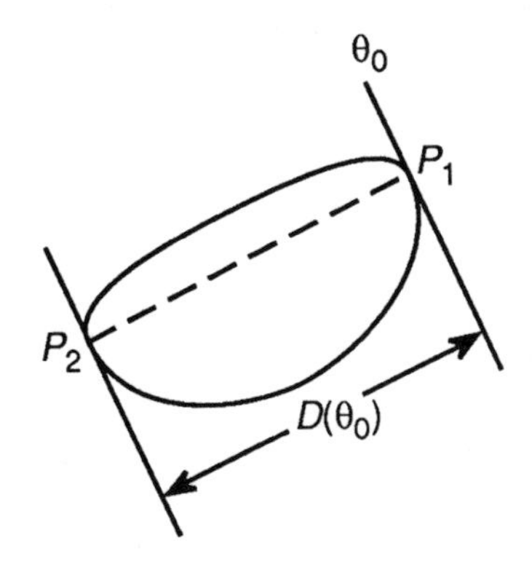

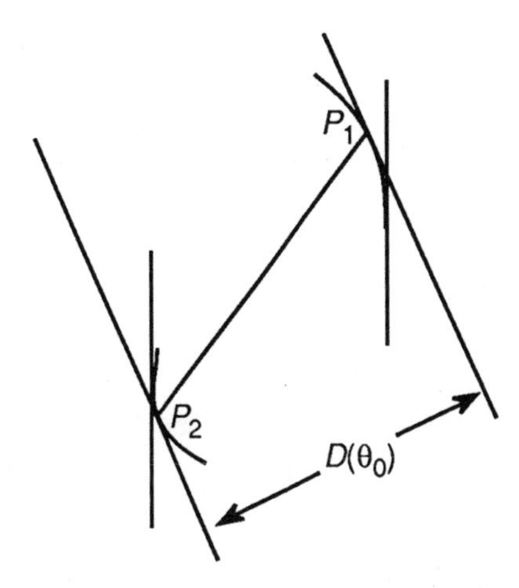

2. If P_1P_2 is not perpendicular to the parallel lines in direction θ_0, then when θ is chosen in the direction perpendicular to P_1P_2 we see $D(\theta) \geq$ length $P_1P_2 > D(\theta_0)$, which contradicts $D(\theta_0)$ being a maximum.

3. The ball repeatedly traces out segment P_1P_2. This is called a periodic billiard path.

THE MEAN VALUE THEOREM

Mean Value Theorem–A Theorem About Parallel Lines

Theorem: If f is continuous on $[a, b]$ and differentiable in (a, b), then there is at least one number $c \in (a, b)$, such that

$$f'(c) = \frac{f(b) - f(a)}{b - a}$$

Exercises on Mean Value Theorem

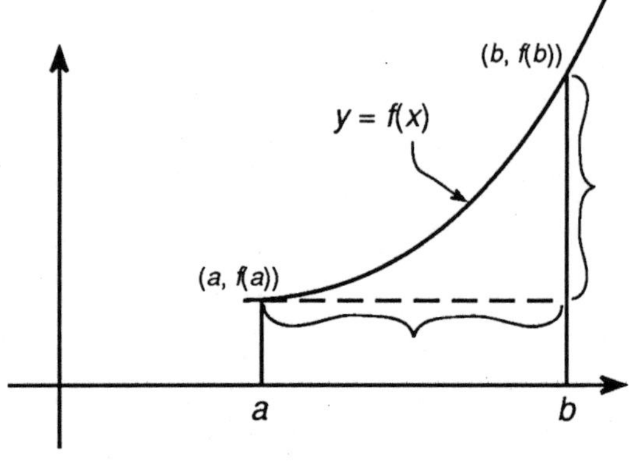

1. In the figure at the right, label a segment whose length is $f(b) - f(a)$.

2. Label a segment whose length is $b - a$.

3. Complete the statement: The quotient $\frac{f(b)-f(a)}{b-a}$ is the ____________ of the segment joining the two points (,) and (,).

4. Complete the statement: $f'(c)$ gives the ____________ of the line ____________ to the curve $y = f(x)$ at the point (,).

5. On the graph shown, locate the "c" that is guaranteed by the Mean Value Theorem and draw in the two parallel lines to which the theorem refers. Thus the conclusion of the theorem can be restated as follows: there will be two parallel lines, one through the points (,) and (,), the other ____________ to the curve $y = f(x)$ at the point (,).

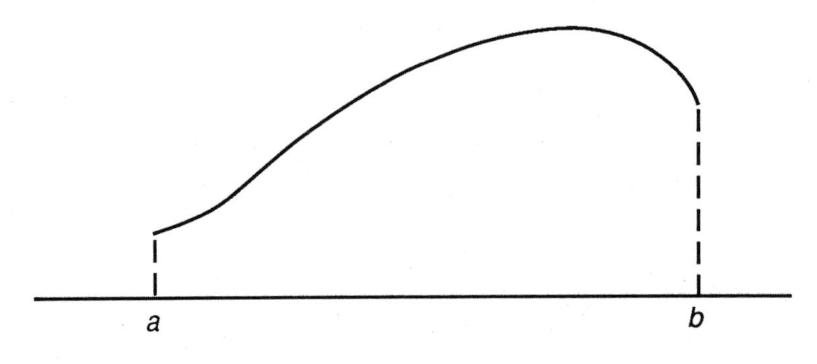

6. What condition could be placed on the graph of f to ensure that at least two such cs can be found in (a, b)?

7. Can a person understand the Mean Value Theorem but not Rolle's Theorem? Explain.

8. *If I drive 80 miles in two hours, at some point along the way my speed was 40 miles per hour.* Justify that statement using the Mean Value Theorem.

Answers to Exercises on Mean Value Theorem

1. and 2.

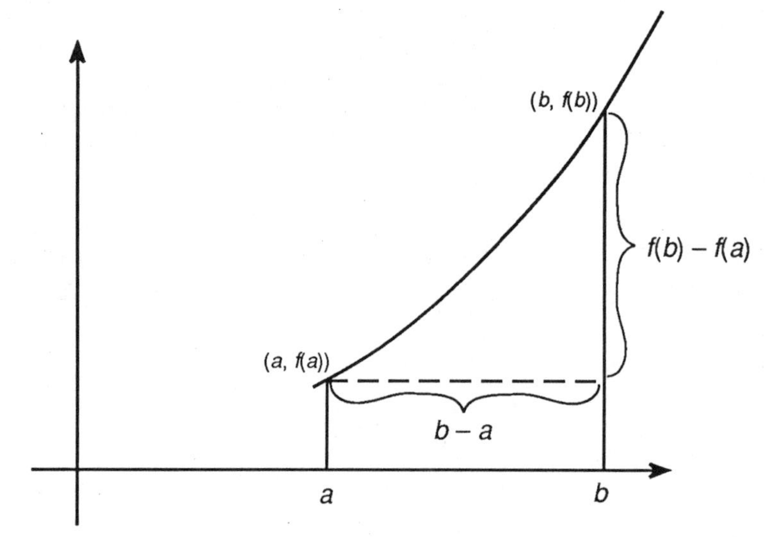

3. slope, $(a, f(a)), (b, f(b))$

4. slope, tangent, $(c, f(c))$

5. Locate c where the tangent line at $(c, f(c))$ is parallel to the line through $(a, f(a))$ and $(b, f(b))$. For this function there is only one such c in $[a, b]$.

6. A number of conditions could be stated. The most obvious is that the graph cross the secant line somewhere between $(a, f(a))$ and $(b, f(b))$ as shown at the right. Then a mean value point c for $[a, b]$ can be found in both the intervals (a, d) and (d, b).

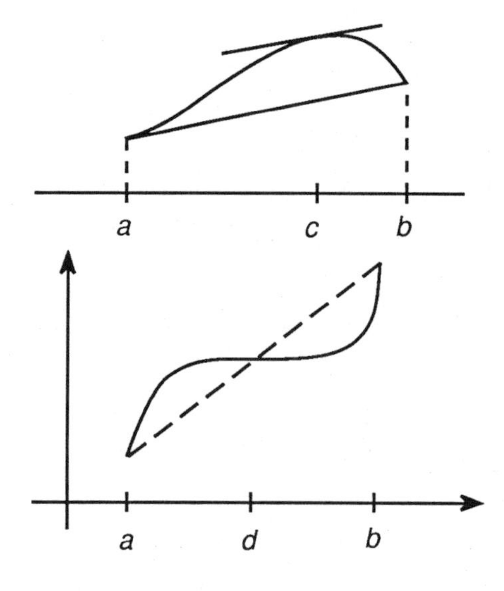

7. Not really. Rolle's Theorem is a special case of the Mean Value Theorem, where one has the additional condition that $f(a) = f(b)$.

8. Let $s(t)$ represent position at time t. For those two hours between t_0 and t_1 we have $s(t_1) - s(t_0) = 80$, $t_1 - t_0 = 2$ (hours). By the Mean Value Theorem (assuming no abrupt stops, so $s'(t)$ exists on (t_0, t_1)) for some $c \in (t_0, t_1)$,

$$s'(c) = \frac{s(t_1) - s(t_0)}{t_1 - t_0} = 40 \text{ mph.}$$

Exercises on Slopes and Speeds

Applications of the Mean Value Theorem

I always drive the 120 miles between home and my Uncle Joe's house along the same route. On my last trip I started (from rest) at 8:00 a.m. At 8:30 I realized I had left my dog Ralph and returned home to get him. I had driven 50 miles total for nothing. I left a second time from home at 9:00 a.m. At 10:00 a.m. I was given a speeding ticket for going 65 mph in a 50 mph zone. I finally pulled into Uncle Joe's driveway at noon.

1. If $s(t)$ gives the miles from home along the trip at time t, and $v(t)$ is my velocity at time t, (positive when moving toward Uncle Joe's, negative when headed toward home), draw a plausible graph for both $s(t)$ and $v(t)$. Decide a suitable scale for both graphs.

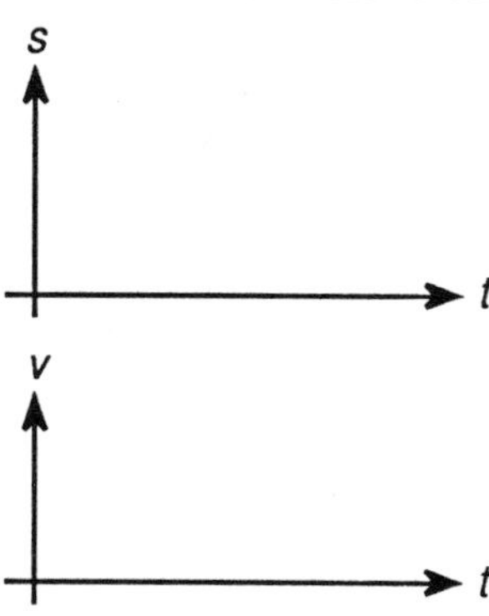

2. The fact that I started and ended the trip at rest is shown on the graph of v by ____________ and on the graph of s by ____________.

3. The fact that I was speeding just before 10:00 a.m. is shown on the graph of v by ____________ and on the graph of s by ____________.

4. For times t where $v(t)$ is largest, the graph of s ____________.

5. For how many different t can you be sure
   (a) $|v(t)| = 40$ (b) $s(t) = 40$

6. What is $\frac{1}{3}\int_9^{12} v(t)dt$? (Skip this problem if you haven't had integrals yet.)

Answers to Exercises on Slopes and Speeds

1.

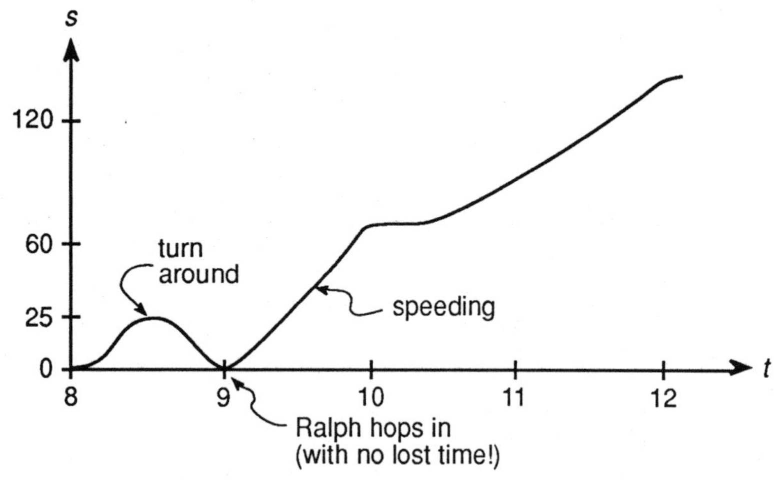

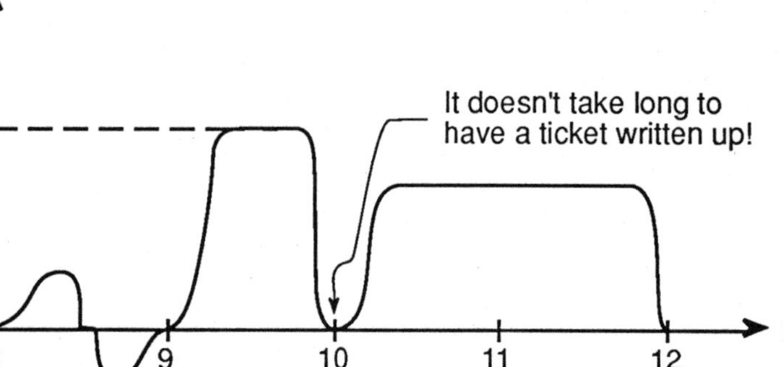

2. v is zero when stopped, s has a horizontal tangent when stopped.

3. v is 65, slope of s is 65 (when appropriately scaled).

4. has largest slope.

5. (a) At least twice between 8 and 9 (to cover 50 miles in one hour, $|v(t)|$ must climb past 40 mph, then fall back through 40 mph); at least twice between 9 and 12. (Average velocity is 40 mph. Starting from rest it was not constantly 40; hence it has to exceed 40 somewhere during this time.)

   (b) Once sometime after 9 o'clock.

6. $\frac{1}{3}\int_9^{12} v(t)dt = \frac{1}{3}(s(12) - s(9)) = \frac{1}{3}(120 - 0) = 40$. This is average velocity over these three hours.

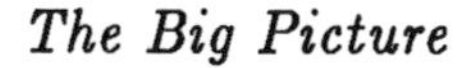

Related Rates

Problems involving "related rates" provide a nice variety of applications using the derivative. We will examine the general procedures that usually follow the pattern outlined below by looking at a common specific example, a ladder that is free to slide down a wall. Assume a 20-foot ladder leans against a wall as shown. Our goal is to calculate the rate at which the bottom of the ladder moves away from the wall at the moment the top is 10 feet from the floor, assuming that the top slides down the wall at 0.5 feet per second.

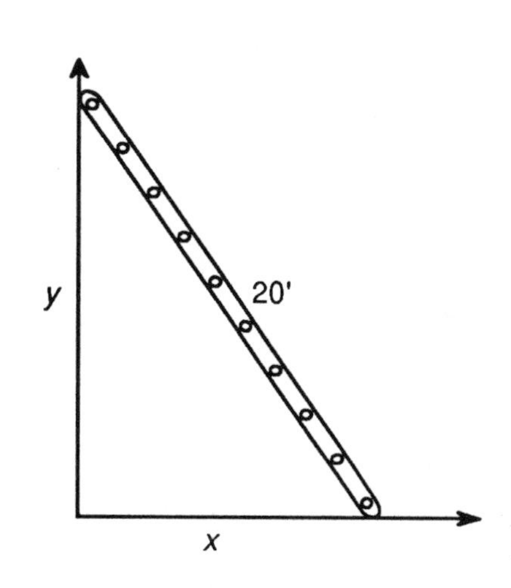

Step 1: Draw a figure and introduce notation for all pertinent variable quantities.

Variable quantities should be denoted by x, y, and so on. Use numerical values (say $20'$) *only* if the quantity is a constant during the motion. A common error is to use $10'$ instead of the variable y; this "freezes" the motion and makes it impossible to deduce rate information.

Step 2: Write the equations that express the relationships among the variable quantities.

In our case, x and y are related because $x^2 + y^2 = 400$. The equation giving the relationship will come from any of a number of formulas, rules, theorems, or laws depending on the problem. Note in our case it is Pythagoras's Theorem that gives the equation.

In *related rates* problems the two words describe the situation. A "rate" is a derivative, a change with respect to time (usually) or some other independent variable. In problems such as ours, we may be told how the ladder is sliding down the wall, that y is decreasing at a certain rate, or, that x is increasing at a certain rate. So "rates" are derivatives and they are "related." That makes sense, because if the quantities themselves are related at all times, then their derivatives (rates) will also be related. The relationship between the rates is easy to find: differentiate the equation that relates the variables themselves. So we have:

Step 3: Differentiate the equation relating the variables with respect to the independent variable on which all variables in the equation depend.

In our case, since $x^2+y^2=400$ and x and y depend on time, we get by differentiation $2x\frac{dx}{dt}+2y\frac{dy}{dt}=0$, or simply $x\frac{dx}{dt}+y\frac{dy}{dt}=0$. That is all there is to Step 3.

Step 4: Substitute the given information into the equations of both Steps 2 and 3.

In our case, we were told that when the top of the ladder is 10 feet off the floor it is sliding down the wall at the rate of .5 feet per second. Thus we are told that when $y=10$, $\frac{dy}{dt}=-.5$ ft/sec.

Step 5: Solve for unknown variables or rates as required by the problem.

When $y=10$ we can solve for x to see that $x=10\sqrt{3}$.

Since $x\frac{dx}{dt}+y\frac{dy}{dt}=0$, we then get $10\sqrt{3}\frac{dx}{dt}-10(.5)=0$ so $\frac{dx}{dt}=\frac{1}{2\sqrt{3}}$ ft/sec.

Caution: A common error students make is to reverse Steps 3 and 4.

In our case, we might have started with $x^2+y^2=400$, inserted $y=10$ to get $x^2+100=400$, and then differentiated to get $2x\frac{dx}{dt}+0=0$ or $x\frac{dx}{dt}=0$. This is impossible since neither x nor $\frac{dx}{dt}$ is 0. When we substitute $y=10$ in $x^2+y^2=400$, we have frozen both x and y; no change (or rate) is possible and differentiation will produce nonsense.

Most related rate problems ask for specific *quantitative* information. Less often we are asked for *qualitative* information. That's too bad because gleaning qualitative information from an equation is an important skill.

What qualitative information do we get from $x\frac{dx}{dt}+y\frac{dy}{dt}=0$? Let's list some consequences.

1. $\frac{dx}{dt}$ and $\frac{dy}{dt}$ always have opposite signs since x and y are non-negative. That makes sense: if x and y were both increasing, the ladder would be stretching.

2. The negative ratio of the rates equals the ratios of the lengths x and y. Since $\frac{x}{y}=\frac{dx/dt}{-dy/dt}$, as the ratio of x/y increases (by the ladder sliding down the wall), the foot of the ladder gains speed (since $\frac{dx/dt}{-dy/dt}$ also has to get bigger).

3. The foot of the ladder is slowing to a stop as the ladder hits the floor, assuming the speed of the ladder top is always less than the speed of light. As $y \to 0$ and $\frac{dy}{dt}$ is bounded, the term $y\frac{dy}{dt}$ goes to zero. Hence it is required that $x\frac{dx}{dt}$ also goes to zero, and since x is approaching 20 that requires $\frac{dx}{dt} \to 0$ as claimed.

Now one problem for practice: A balloon is blown up at a constant rate of 100 in^3/min. How fast is the radius increasing when the sphere contains 50 in^3 of gas? How fast is the surface area increasing at that time?

Step 1: Let V, S, and r denote the respective volume, surface area, and radius of the spherical balloon.

Step 2: Write down relationships among variables: $V = (4/3)\pi r^3, S = 4\pi r^2$.

Step 3: Differentiate $\frac{dV}{dt} = 4\pi r^2 \frac{dr}{dt}$ and $\frac{dS}{dt} = 8\pi r\frac{dr}{dt}$.

Step 4: Plug in given values. For us $V = 50$ in^3 and $\frac{dV}{dt} = 100$ in^3/min. So we have $50 = (4/3)\pi r^3$, and thus $r = \sqrt[3]{\frac{150}{4\pi}} = \sqrt[3]{\frac{75}{2\pi}}$ when $V = 50$. Also from $\frac{dV}{dt} = 4\pi r^2 \frac{dr}{dt}$ we have $100 = 4\pi r^2 \frac{dr}{dt}$.

Step 5: From $100 = 4\pi r^2 \frac{dr}{dt}$ and $r = \sqrt[3]{\frac{75}{2\pi}}$ we solve for $\frac{dr}{dt}$ and get

$$100 = 4\pi \left(\sqrt[3]{\frac{75}{2\pi}}\right)^2 \frac{dr}{dt},$$

$$\text{so} \quad \frac{dr}{dt} = \left(\frac{2\pi}{75}\right)^{2/3} \cdot \frac{100}{4\pi} \text{ in/min} \approx 1.5 \text{ in/min}.$$

Also $\frac{dS}{dt} = 8\pi r\frac{dr}{dt}$. We now know r and $\frac{dr}{dt}$ when $V = 50$ in^3. These can be plugged in to get $\frac{dS}{dt} = 200\left(\frac{2\pi}{75}\right)^{\frac{1}{3}} \approx 87.5$ in^2/min.

Exercises on Related Rates

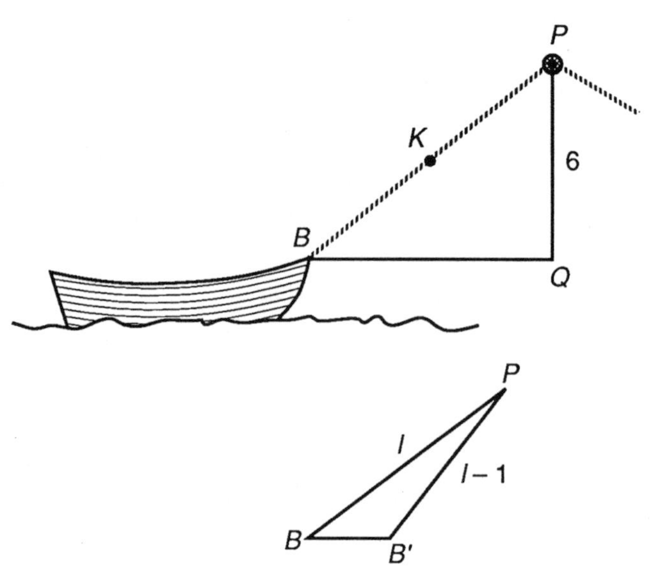

1. A boat is being drawn toward shore by a rope over a pulley 6 feet above boat level. A knot is in the rope at point K as shown. If one foot of rope is retrieved (so KP is reduced by one unit), is the distance BQ reduced by more or less than one foot?

2. Question 1 can be answered without using calculus. Use the diagram to the right to give the answer. What inequality is used?

3. The top of a ladder slides down a vertical wall at a constant speed. When is the base of the ladder moving away from the wall

   (a) at the same speed the ladder is sliding down the wall?

   (b) twice as fast as the ladder is sliding down the wall?

   (c) the fastest?

   (d) If the base of the ladder moves away from the wall at a constant rate and you are on the top rung, what will happen to you when you hit the floor?

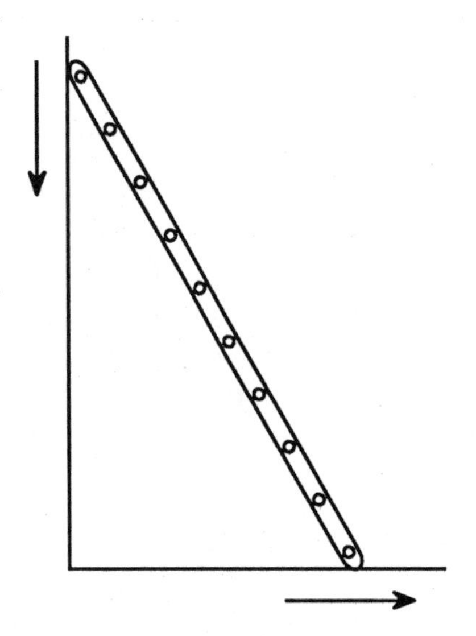

4. A spherical balloon is being inflated so that the surface area of the sphere increases at a constant rate. So $\frac{dS}{dt} = m$ or $S = mt$. Both the radius and volume of the sphere are also increasing. Sketch what you think are reasonable graphs for the radius $r(t)$ and volume $V(t)$ as the surface area increases linearly. Can you prove your graphs are reasonable?

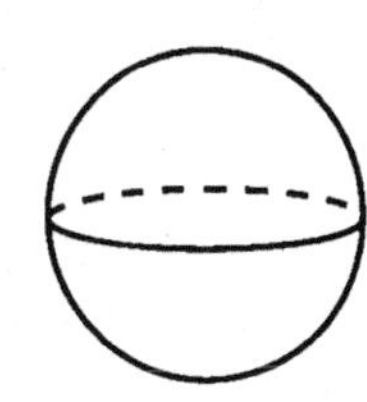

5. *A Surprise!* The formula $C = 2\pi r$ for the circumference of a circle gives $\frac{dC}{dt} = 2\pi\frac{dr}{dt}$. So if $\frac{dr}{dt} = 1$, in one second the radius of the circle increases one unit and the circumference increases by $2\pi \approx 6.28$ units *regardless of the size of the circle.* A one-unit increase in radius always corresponds to a 2π-unit increase in circumference and vice versa. The surprising thing is that the size of the circle has nothing to do with our observation. Thus if a 2π-foot length of string is spliced into a string tightly wound around a dime and a new circle, formed with the same center, the new circle clears the dime by one foot at every point. If the same 2π-length of string is spliced into one tightly wound around the equator of the earth, the new circle will clear the earth by one foot at every point just as it did for the dime.

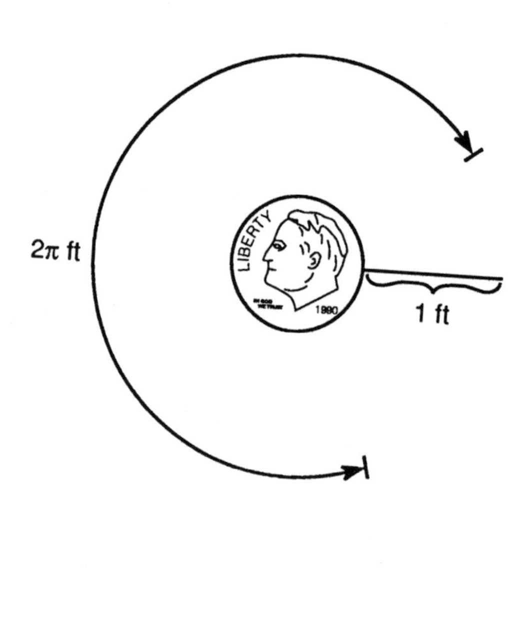

Acknowledgment: Problems 1 and 2 are due to Bill Leonard, Professor of Mathematics at California State University, Fullerton.

Answers to Exercises on Related Rates

1. Let $BQ = x$, so that $\ell^2 = x^2 + 6^2$. Differentiation gives the rate equation $\ell\frac{dl}{dt} = x\frac{dx}{dt}$. Since $\ell > x$, it follows that $|\frac{d\ell}{dt}| < |\frac{dx}{dt}|$, which means that the distance x decreases more rapidly than the length ℓ of the rope. During the time ℓ decreases by a foot, x will decrease by more than one foot.

2. By the triangle inequality, the length BB' is greater than 1.

3. From $x^2 + y^2 = a^2$ we have
   $\frac{dx}{dt} = -\left(\frac{y}{x}\right)\frac{dy}{dt}$ and we are given that $\frac{dy}{dt}$ is constant.

   (a) when $x = y$

   (b) when $\frac{y}{x} = 2$

   (c) when x is near zero

   (d) you would break your back. Since $\frac{dx}{dt}$ is constant and $\frac{y}{x}$ approaches zero $\mid \frac{dy}{dt} \mid$ becomes unbounded.

4. For large balloons ($r > 1$), the surface area grows faster than the radius, the volume faster than surface area. This follows from the general fact that if r is replaced by λr (e.g., r is multiplied by 2), surface area is multiplied by λ^2 and volume by λ^3. Specifically note that $kt = S = 4\pi r^2$ so $r = \sqrt{\frac{k}{4\pi}t}$; then $V = \frac{4}{3}\pi r^3$ gives $V = \frac{4}{3}\pi(\frac{k}{4\pi}t)^{3/2} = \frac{1}{3\sqrt{4\pi}}(kt)^{3/2}$.

5. This is one of those problems that is interesting because it is counterintuitive.

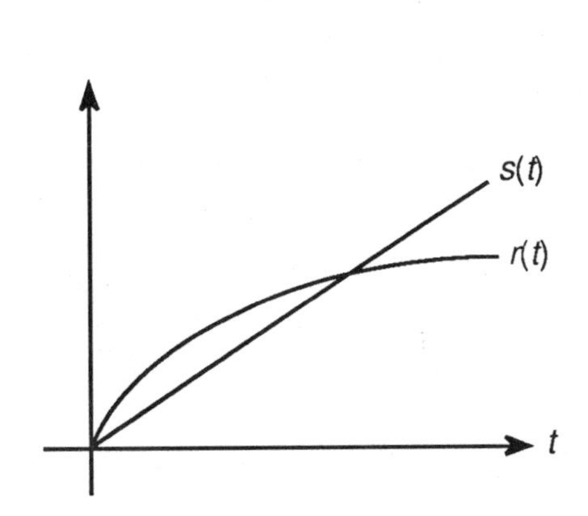

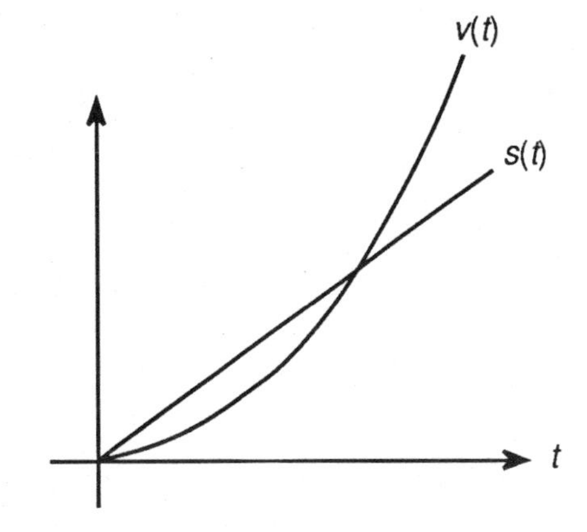

Exercises on Vessels and Volumes

More Related Rates

1. For the vessel shown at the right, plot a reasonable curve for volume V as a function of height h.

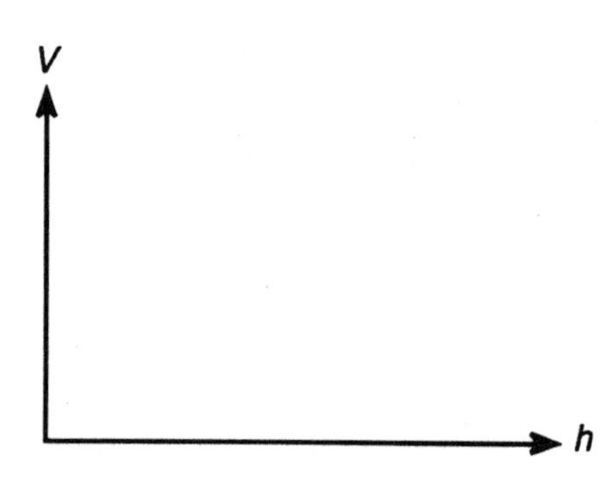

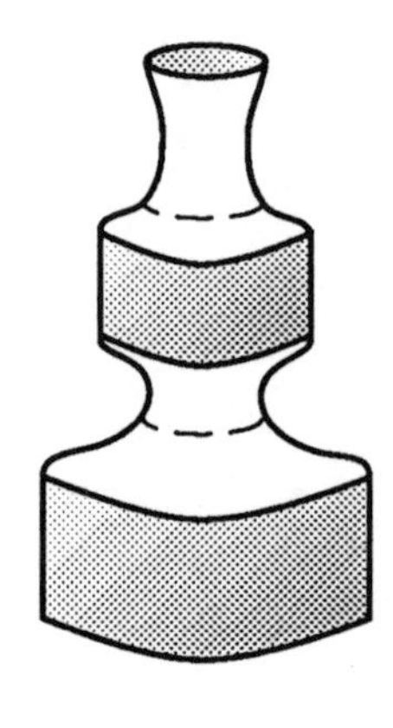

2. For the vessel shown in Exercise 1, water is poured in so that dh/dt is constant. Plot a reasonable curve for h, V, and dV/dt as functions of time.

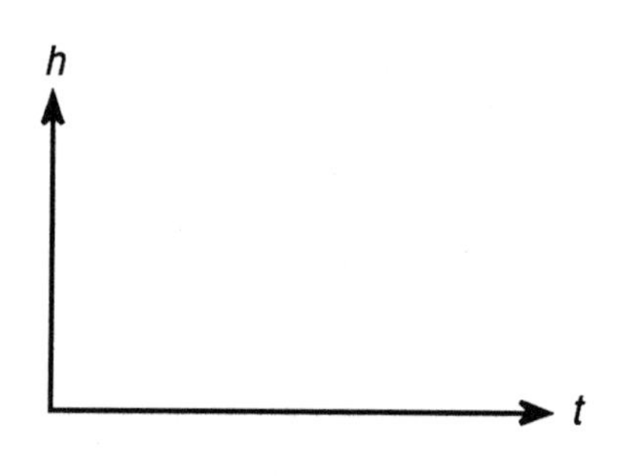

V

t

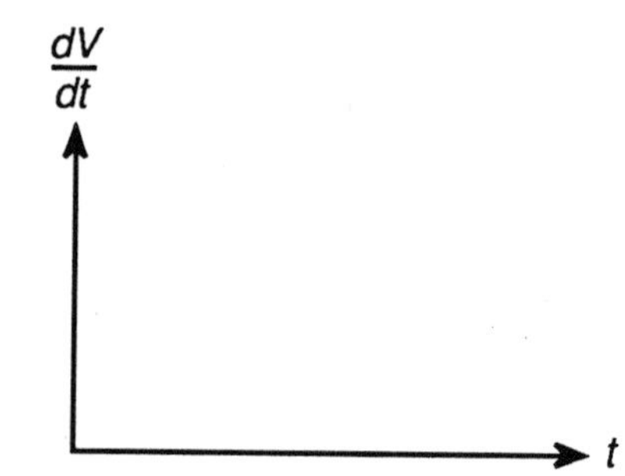

3. Water is poured into the vessel at a constant rate (that is, $\frac{dV}{dt}$ is constant.) Plot a reasonable curve of h and dh/dt as functions of time.

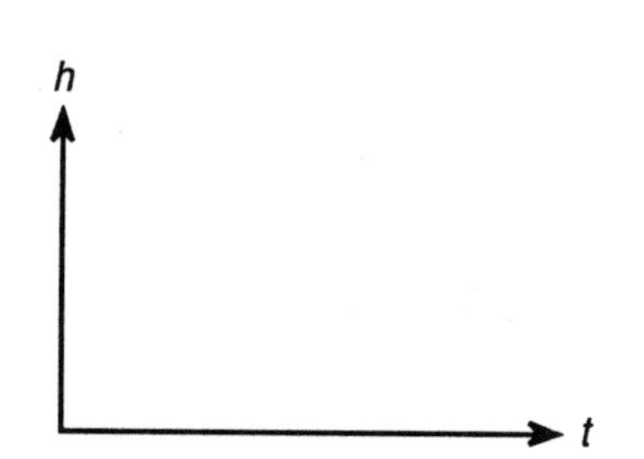

$\frac{dh}{dt}$

t

4. The graph given shows the volume in a vessel as a function of height of water in the vessel. Draw a picture of the vessel.

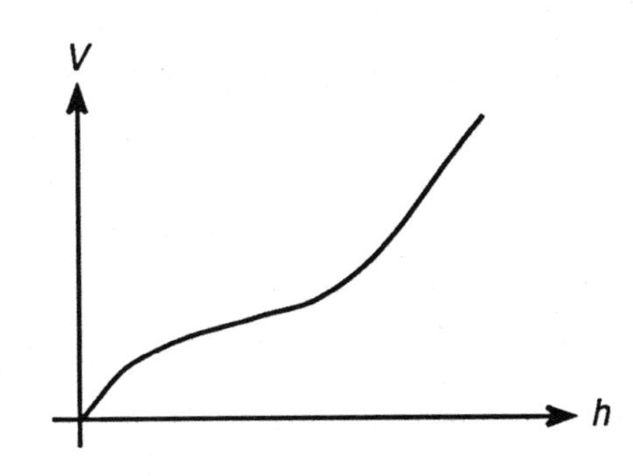

5. Water is poured into a vessel at a constant rate. The graph of the height of the water in the vessel, as a function of time, is shown. Draw a picture of the vessel.

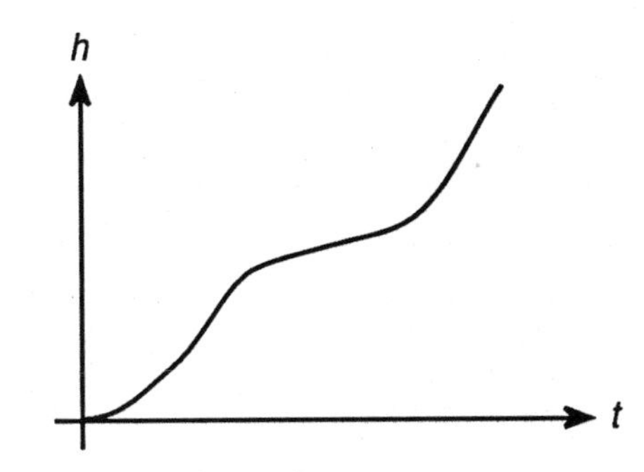

Answers to Exercises on Vessels and Volumes

1. Note that the volume, as a function of height, increases rapidly when the vessel is large in cross section. The increase in volume is at a constant rate where the vessel has vertical sides–this corresponds to straight line segments on the graph.

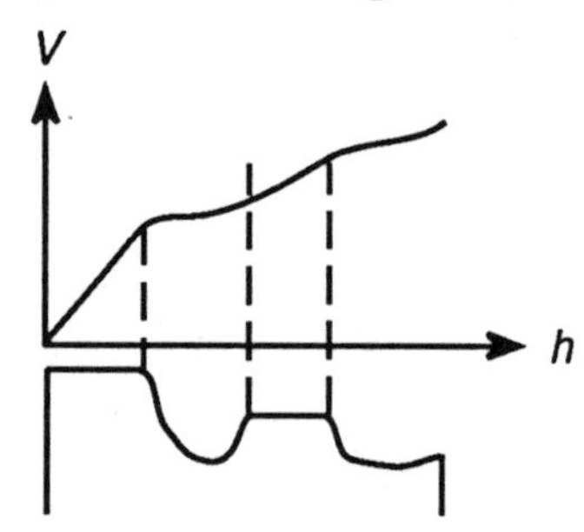

2. Note that the linear segments on the graph of V correspond to the water rising at a constant rate in the vessel with vertical sides. The first linear segment has greater slope because volume increases faster there than it does where the vessel has smaller cross section. Note also, $\frac{dV}{dt}$ is constant when the part of the vessel with vertical sides is filling. Also, $\frac{dV}{dt}$ is largest where cross sections are largest. Note the similar shapes of the graph of V as a function of time here and the graph of V as a function of height in problem 1. This is explained by $\frac{dh}{dt} = k$, a constant.

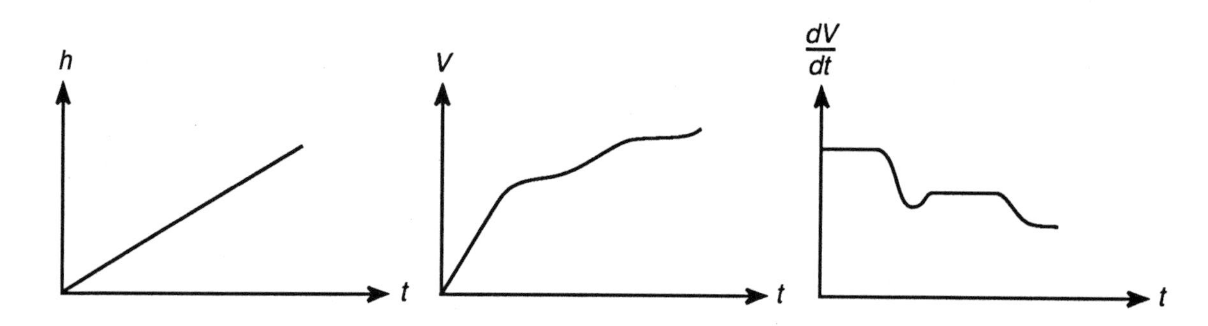

3. Note that the linear segments on the graph of h correspond to the vertical sides of the vessel. The second linear segment has greater slope because h increases faster where the vessel's cross section has smaller area. Also note, $\frac{dh}{dt}$ is greatest where the cross sections are smallest.

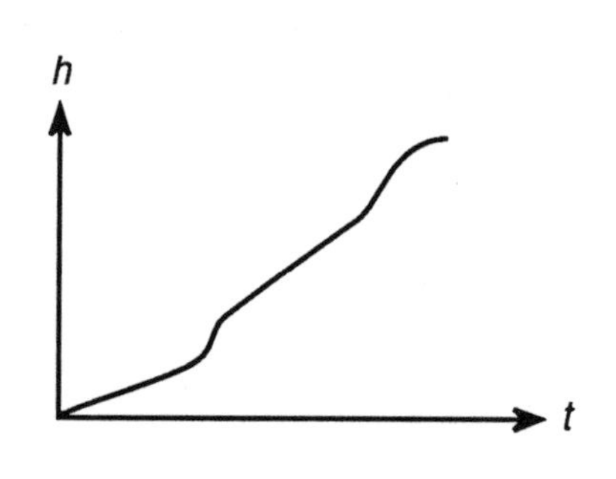

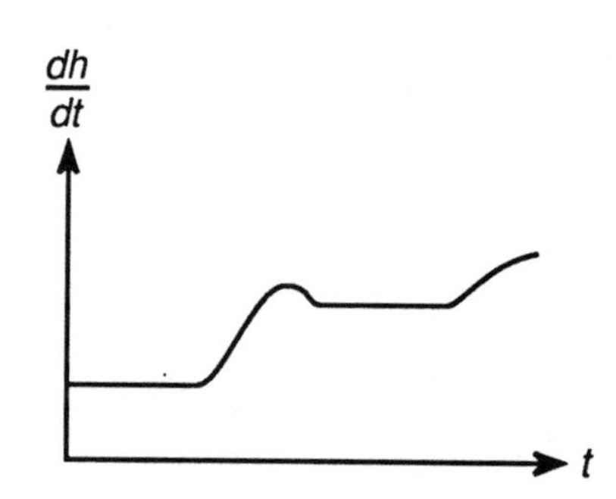

4.

5.

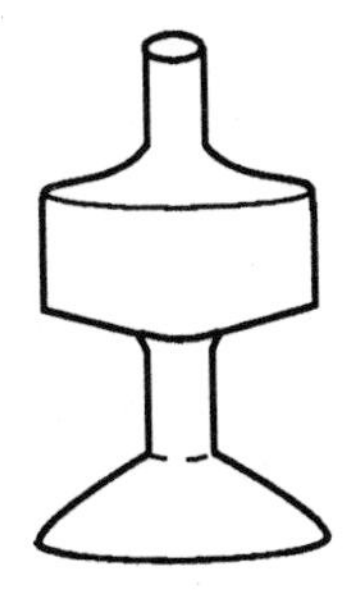

Comment: Students may note that the graph of $\frac{dV}{dt}$ in problem 2 looks like the vessel laid on its side. The similarity in shape is explained as follows. Assume $h = kt$ since we are given that $\frac{dh}{dt}$ is constant. If the upright vessel has radius $r(h)$ at height h, then $V(h) = \int_0^h \pi(r(h))^2 dh$ so that $\frac{dV}{dt} = \frac{dV}{dh} \cdot \frac{dh}{dt} = \pi(r(h))^2 \cdot k$. Thus the value of $\frac{dV}{dt}$ is proportional to the square of the radius of the vessel.

Interpreting Related Rates Equations

In applied mathematics, the variables that describe a real-world problem will have physical or geometrical meaning. Typical symbols x, t, V, P, R may therefore represent distance, time, volume, pressure, and revenue, and their derivatives express the rate of change of these quantities. It is an important practical skill, and often an interesting challenge, to interpret what the equations among the variables and derivatives say about the real-world situation that is being modeled.

Related rates problems provide good practice in interpreting information from equations. In what follows, we look at two problems and their solutions, and then discuss what can be learned from the equations derived in the course of solution.

Problem 1. Water is being pumped at a constant but unknown rate from a conical tank 10 meters high and 8 meters across. At noon the height of water is 7 meters, and you observe that the water level drops 3 cm over the next minute. Estimate when the tank will be pumped empty.

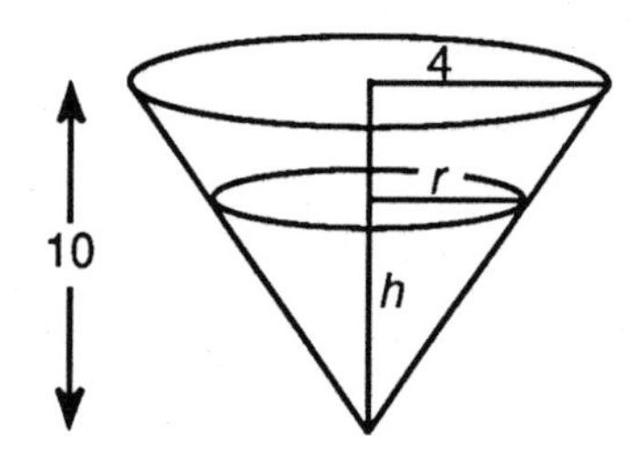

Solution. By similar triangles, when the water level is h meters above the bottom of the tank, the water surface is a circle of radius

$$r = \left(\frac{h}{10}\right) 4\ m. \tag{1}$$

Thus the volume of water in the tank, V, is given by

$$V = \frac{1}{3}(\pi r^2)h = \frac{4\pi}{75}h^3\ m^3. \tag{2}$$

If k denotes the rate at which water is pumped from the tank, then $\frac{dV}{dt} = -k$. Differentiating (2), we find

$$\frac{dV}{dt} = -k = \frac{12\pi}{75}h^2\frac{dh}{dt}. \tag{3}$$

At noon, we observe that $h_0 = 7\ m$ and, at least approximately,

$$\left|\frac{dh}{dt}\right|_0 \approx 3\frac{\text{cm}}{\text{min}} = 1.8\ \frac{\text{m}}{\text{hr}}$$

Thus

$$-k \approx \frac{12}{75}\pi(7)^2(-1.8)\ m^3/hr$$

or

$$k \approx 44.3\ m^3/hr.$$

At noon, the tank held

$$V_0 = \frac{4\pi}{75}(7)^3\ m^3 \approx 57.5\ m^3,$$

and so the tank empties at $V_0/k = (57.5\ m^3)/(44.3\ m^3/hr) \approx 1.3$ hours past noon; that is, at about 1:18 p.m. the tank should be empty.

Discussion. Since k is a rate at which water is *removed* from the tank, the rate of change of volume is negative. This was reflected in equation (3) by the negative sign preceding the k. Also from (3) we see that $h^2\frac{dh}{dt}$ is a negative constant and so $\frac{dh}{dt} < 0$, which of course means that h is decreasing. Also, as h and therefore h^2 decreases, we see that $\left|\frac{dh}{dt}\right|$ increases. That is, the water level drops at increasing speed as the pumping continues.

In their present form, equations (1), (2), and (3) have been algebraically processed in a way that reaches the solution in an efficient way. Nevertheless, there is an alternative way to arrange the equations that reveals an important principle at work. Let's express V in terms of r instead of as a function of h. That is, using (1),

$$V = \frac{1}{3}(\pi r^2)h = \frac{1}{3}\left(\pi r^2\right)\left(\frac{10}{4}r\right) = \left(\frac{10}{4}\right)\left(\frac{\pi}{3}r^3\right).$$

Then

$$\frac{dV}{dt} = \left(\pi r^2\right)\left(\frac{10}{4}\frac{dr}{dt}\right) = \left(\pi r^2\right)\frac{dh}{dt}. \tag{4}$$

Here we recognize that πr^2 is the area of the water surface, and so

$$\frac{dV}{dt} = A\frac{dh}{dt}, \tag{5}$$

where A is the surface area of the water at height h. It is not difficult to reason that (5) is a correct formula for tanks of *any* shape, not just conical tanks. We can now see from (5) that the rate at which the water level changes is directly proportional to the pumping rate and inversely proportional to the cross-sectional surface area of the tank.

Problem 2. The makers of the adventure film *Washington Jones* wish to film the hero's jump off a 95-foot cliff into a crocodile-infested pool. A camera is set up 30 feet from the cliff at a height of 70 feet, and will follow the fall of the stunt actor. Past experience shows that the camera can track at angular rates of no more than 60 degrees per second. How far into the fall will the camera track the action? or, can the entire fall be recorded on film?

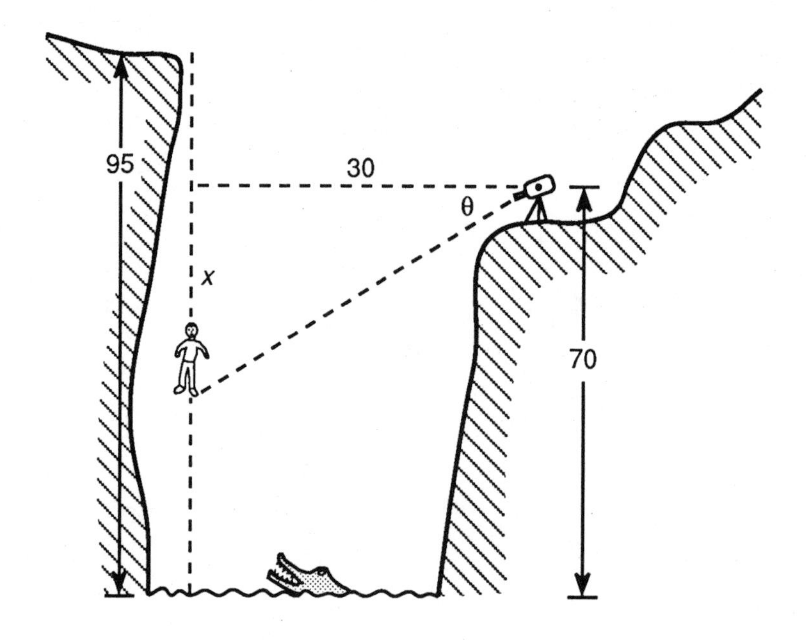

Solution. If x denotes the height of the stunt actor below camera level, then $25 + x$ is the distance of fall. Assuming we can neglect air resistance, and taking $g = 32 ft/\sec^2$ as the gravitational acceleration, we have

$$25 + x = 16t^2. \tag{1}$$

Differentiating (1), we get

$$\frac{dx}{dt} = 32t. \tag{2}$$

To bring in θ, we observe that

$$\tan\theta = \frac{x}{30}, \quad x \geq -25, \tag{3}$$

and so

$$(\sec^2\theta)\frac{d\theta}{dt} = \frac{1}{30}\frac{dx}{dt}. \tag{4}$$

We now wish to solve for the value of x when $\frac{d\theta}{dt}$ reaches $\pi/3$ radians per second, the angular speed in radian measure that corresponds to $60°$/second.

Solving for t in equation (1) and inserting this expression in (2), we find

$$\frac{dx}{dt} = 32t = 32\left(\frac{1}{4}\sqrt{25+x}\right) = 8\sqrt{25+x}. \tag{5}$$

Using (3) and a trigonometric identity we also have

$$\sec^2\theta = 1 + \tan^2\theta = 1 + \left(\frac{x}{30}\right)^2. \tag{6}$$

When (5) and (6) are put into (4) we get the following equation for x:

$$\left[1 + \left(\frac{x}{30}\right)^2\right]\left(\frac{\pi}{3}\right) = \frac{8}{30}\left[\sqrt{25+x}\right], \tag{7}$$

where $x \geq -25$.

The answer to the question posed in the problem is now reduced to finding the smallest value of x that satisfies (7), or showing that (7) has no solution. While a variety of methods are available, perhaps the easiest approach is to graph the left- and right-hand sides of (7) using the graphics capabilities of a computer or calculator. The graphs below show that the curves first cross at about $(-7.6, 1.1)$. Thus the camera can follow the fall only from $x = -25$ to $x = -7.6$.

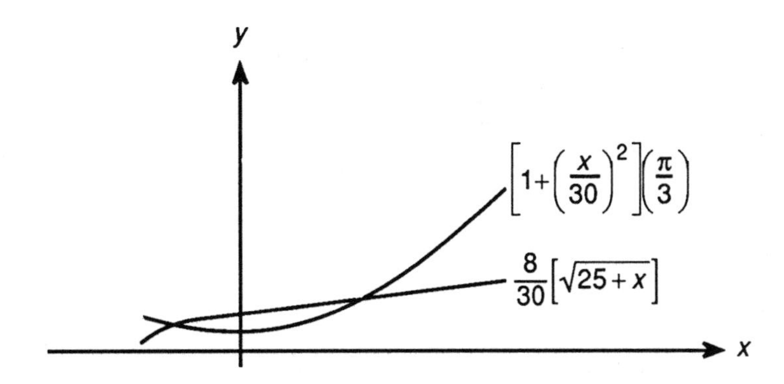

Discussion. The two curves graphed above are parabolas whose exact placement depends on the specified values given in the problem. If the camera were in a different location relative to the top of the cliff, the parabolas may not intersect. Of special interest is how the angular rate $\frac{d\theta}{dt}$ alters the position of the vertical parabola. As equation (7) reveals, $\frac{d\theta}{dt} = \frac{\pi}{3}$ is a scaling factor that raises and lowers the upper parabola as $\frac{d\theta}{dt}$ is respectively increased or decreased. This is sensible on physical grounds. If $\frac{d\theta}{dt}$ is small, the parabolas intersect and there is an interval of x-values in which θ changes too rapidly for the camera to follow. On the other hand, for large enough $\frac{d\theta}{dt}$ the parabolas are nonintersecting and the camera can track the entire motion. This is shown in the following graph, where we assume $\frac{d\theta}{dt}$ is as large as 120° per second. With this assumption the factor $\frac{\pi}{3}$ in equation (7) is replaced by $\frac{2\pi}{3}$.

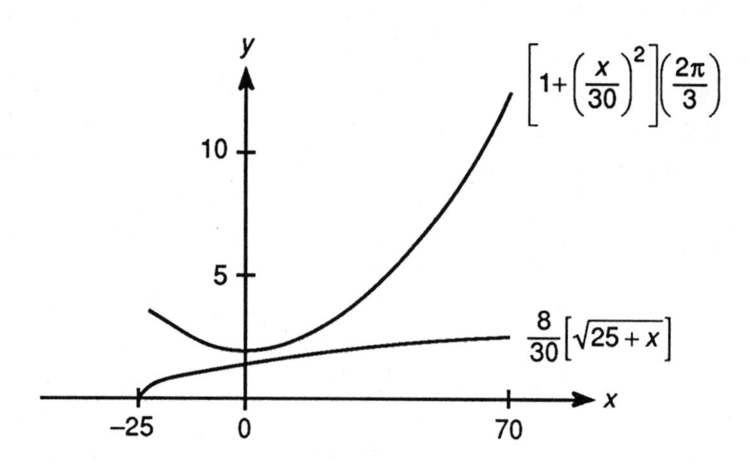

Projects

1. There is enough information in Problem 1 to compute, without calculus, the exact amount of water pumped out during the one minute it took for the height h to drop from 7 meters to 6.97 meters. Show directly that $k = 44.14$ cubic meters per hour, without using equation (3).

2. The camera in Problem 2 is at a distance b from the line of fall, so that b replaces the 30s that appear in the equations. Discuss the effect of large and small b values.

3. Take any related rates problem and its solution and go one more step: provide a discussion of how the variables and parameters interact, using the discussions above as models.

The Integral of sin x via Riemann Sums

The definite integral of a function defined on an interval $[a, b]$ is defined as the limit of Riemann sums. It is possible (but not all that easy) to compute the integrals of $f(x) = x^n$ by considering Riemann sums, but most functions lead to Riemann sums whose limits cannot be easily evaluated. Here we show that a trigonometric identity allows the integral of $\sin x$ to be evaluated.

By setting $A = k\theta$ and $B = \frac{1}{2}\theta$ in the identity $2\sin A \sin B = \cos(A-B) - \cos(A+B)$, we find

$$2\sin(k\theta)\sin(\frac{1}{2}\theta) = \cos[(k-\frac{1}{2})\theta] - \cos[(k+\frac{1}{2})\theta]. \tag{1}$$

Choosing $k = 1, \dots, n$ in (1), the resulting n equations are:

$$\begin{cases} 2\sin(\theta)\sin(\frac{1}{2}\theta) &= \cos(\frac{1}{2}\theta) - \cos(\frac{3}{2}\theta) \\ 2\sin(2\theta)\sin(\frac{1}{2}\theta) &= \cos(\frac{3}{2}\theta) - \cos(\frac{5}{2}\theta) \\ \dots & \\ 2\sin(n\theta)\sin(\frac{1}{2}\theta) &= \cos[(n-\frac{1}{2})\theta] - \cos[(n+\frac{1}{2})\theta]. \end{cases} \tag{2}$$

When these n equations are summed, all but two terms on the right side cancel,* and we get

$$2\sin(\frac{1}{2}\theta)\sum_{k=1}^{n}\sin(k\theta) = \cos(\frac{1}{2}\theta) - \cos[(n+\frac{1}{2})\theta]. \tag{3}$$

To evaluate $\int_0^b \sin x dx, b > 0$ we use the partition $\{0, \frac{b}{n}, \frac{2b}{n}, \dots, \frac{kb}{n}, \dots, b\}$ of $[0, b]$ and form the Riemann sum

$$R_n = \sum_{k=1}^{n}\sin(\frac{kb}{n})\ \frac{b}{n}. \tag{4}$$

This formula suggests that we choose $\theta = \frac{b}{n}$ in identity (3), so that it becomes

$$2\sin(\frac{b}{2n})\sum_{k=1}^{n}\sin(\frac{kb}{n}) = \cos(\frac{b}{2n}) - \cos[(n+\frac{1}{2})\frac{b}{n}]. \tag{5}$$

Eventually we will let $n \longrightarrow \infty$, so we may as well assume n is so large that $\frac{b}{2n} < \pi$. This ensures us that $\sin(\frac{b}{2n}) \neq 0$, and so we can multiply both sides of (5) by $(\frac{b}{2n})/\sin(\frac{b}{2n})$, obtaining

$$(\frac{b}{n})\sum_{k=1}^{n}\sin(\frac{kb}{n}) = [\cos(\frac{b}{2n}) - \cos(b+\frac{b}{2n})]\left[\frac{(\frac{b}{2n})}{\sin(\frac{b}{2n})}\right]. \tag{6}$$

*Sums like these are called "telescoping sums."

That is,

$$R_n = [\cos(\frac{b}{2n}) - \cos(b + \frac{b}{2n})] \left[\frac{(\frac{b}{2n})}{\sin(\frac{b}{2n})} \right]. \tag{7}$$

Since

$$\lim_{n\to\infty} \left[\frac{\frac{b}{2n}}{\sin(\frac{b}{2n})} \right] = \lim_{t\to 0} \left[\frac{t}{\sin t} \right] = 1 \ ,$$

we see from (7) that

$$\lim_{n\to\infty} R_n = \cos 0 - \cos b.$$

Thus we have derived the formula

$$\int_0^b \sin x dx = 1 - \cos b. \tag{8}$$

It then follows that

$$\int_a^b \sin x \ dx = \int_0^b \sin x \ dx - \int_0^a \sin x \ dx = -\cos b + \cos a. \tag{9}$$

Exercise on Riemann Sums

Look up an appropriate trigonometric identity that allows you to integrate $\cos x$ by Riemann sums.

Answer to Exercise on Riemann Sums

The relevant equations to integrate $\cos x$ by Riemann sums are these:

$$\begin{aligned}
2\cos A \sin B &= \sin(A+B) - \sin(A-B) \\
2\cos(k\theta)\sin(\tfrac{1}{2}\theta) &= \sin[(k+\tfrac{1}{2})\theta] - \sin[(k-\tfrac{1}{2})\theta] \\
2\sin\tfrac{1}{2}\theta \sum_{k=1}^{n} \cos k\theta &= \sin[(n+\tfrac{1}{2})\theta] - \sin\tfrac{1}{2}\theta \\
R_n = \sum_{k=1}^{n} \cos(\frac{kb}{n})\frac{b}{n} &= \frac{b}{2n}\,\frac{\sin[(n+\frac{1}{2})\frac{b}{n}] - \sin\frac{b}{2n}}{\sin\frac{b}{2n}} \\
\lim_{n\to\infty} R_n = \sin b &= \int_0^b \cos x \; dx. \\
\int_a^b \cos x \; dx &= \sin b - \sin a
\end{aligned}$$

Discovering Properties of $\ln x$ via Riemann Sums

The natural logarithm function $\ln x$ can be introduced in two ways, either as the inverse function of $\exp x$ or as an integral $\int_1^x \frac{1}{t}dt$. If the inverse function definition is adopted, the rules of $\ln x$ are derived on the basis of corresponding rules of $\exp x$. Suppose, however, that the integral definition is taken: can the rules of $\ln x$ still be discovered? Appropriate manipulations of Riemann sums show us the answer is yes. Notice that we do not use the derivative or the fundamental theorem of calculus.

To begin, let's consider $\ln b - \ln a = \int_a^b \frac{1}{x}dx$, where $0 < a < b$. By introducing any partition $P_n[a,b] = \{a = x_0, x_1, \ldots, x_n = b\}$, we consider the Riemann sum

$$R_n[a,b] = \sum_{k=1}^{n} \frac{1}{x_k}[x_k - x_{k-1}].$$

If $\delta(P_n) = \max|x_k - x_{k-1}|, k = 1, \ldots, n$, then we know $R_n[a,b] \to \ln b - \ln a$ as $\delta(P_n) \to 0$.

Next, let $t = b/a$ and define $\hat{P}_n[1,t] = \{\frac{x_0}{a}, \frac{x_1}{a}, \ldots, \frac{x_n}{a}\}$. We see that this is a partition of $[1,t]$, since $\frac{x_0}{a} = \frac{a}{a} = 1$ and $\frac{x_n}{a} = \frac{b}{a} = t$. Now

$$\begin{aligned} \hat{R}_n[1,t] &= \sum_{k=1}^{n} \frac{1}{(x_k/a)}[\frac{x_k}{a} - \frac{x_{k-1}}{a}] \\ &= \sum_{k=1}^{n} \frac{1}{x_k}[x_k - x_{k-1}] \\ &= R_n[a,b]. \end{aligned}$$

Since $\hat{R}_n[1,t] \to \int_1^t \frac{1}{x}dx = \ln t = \ln \frac{b}{a}$ as $\delta(\hat{P}_n) = \frac{1}{a}\delta(P_n) \to 0$ we obtain the result

(1) $\ln b - \ln a = \ln \frac{b}{a}$.

We also know that $\int_1^1 \frac{dx}{x} = 0$; that is,

(2) $\ln 1 = 0$.

Therefore, by taking $b = 1$ in identity (1) we find

(3) $-\ln a = \ln \frac{1}{a}$.

Next we can replace b with ab in identity (1), obtaining

$$\ln ab - \ln a = \ln \frac{ab}{a}.$$

This rearranges to become

(4) $\ln ab = \ln a + \ln b$.

It is then easy to check that (1), (2), (3), (4) hold for *all* $a, b > 0$ (not just for $0 < a < b$).
From (4) it follows easily by induction that

$$\ln a^n = \ln(a \cdot a \cdots a) = \ln a + \cdots + \ln a = n \ln a.$$

Replacing a by $a^{1/n}$, we obtain

$$\ln a = \ln(a^{1/n})^n = n \ln a^{1/n};$$

that is, $\ln a^{1/n} = \frac{1}{n} \ln a$.

Thus $\ln a^{(m/n)} = \ln(a^{1/n})^m = m \ln(a^{1/n}) = (\frac{m}{n}) \ln a$, and so we conclude that

(5) $\ln a^r = r \ln a$

for all positive rational numbers $r = \frac{m}{n}$. Indeed, from (3), it holds for all rational numbers, both positive and negative.

Comments on Discovering Properties of $\ln x$ via Riemann Sums

Since $\ln 2 > 0$, we see that $\ln 2^n = n \ln 2$ for all $n, -\infty < n < \infty$, which shows the range of $\ln x$ is unbounded from below and above. Since $\frac{1}{t} > 0$, we conclude $y = \int_1^x \frac{1}{t} dt = \ln x$ is strictly increasing on $x > 0$, with range $(-\infty, \infty)$. Thus $y = \ln x$ has an inverse function, which we call $\exp y$.

Let us bring in the derivative. Clearly, $\frac{d}{dx} \ln x = \frac{d}{dx} \int_1^x \frac{1}{t} dt = \frac{1}{t}$ from the fundamental theorem of calculus.

From what we know about derivatives of inverse functions, we see that

$$\frac{d}{dy} \exp y = \frac{1}{(\ln x)'} = \frac{1}{(1/x)} = x = \exp y.$$

Of special interest is the number (call it e) for which $\ln e = 1$. Letting $a = e$ in (5), we have

(6) $\ln e^x = x \ln e = x.$

This tells us that

(7) $\exp x = e^x,$

which pretty well completes the story of investigating the logarithm-exponential pair of functions by beginning with the logarithm defined as the integral $\int_1^x \frac{1}{t} dt$.

Notice that "a^b" is a meaningful expression for all real numbers $a > 0$ and b once $\ln x$ and its inverse function $\exp x$ have been introduced, since we can define $a^b = \exp(b \ln a)$.

Probes

Antidifferentiation

We know that any two antiderivatives (indefinite integrals) for the same function on an interval differ by a constant.

1. Suppose $F' = f$ where the graph of F is shown to the right. Suppose $G' = f$. Draw a possible graph for G.

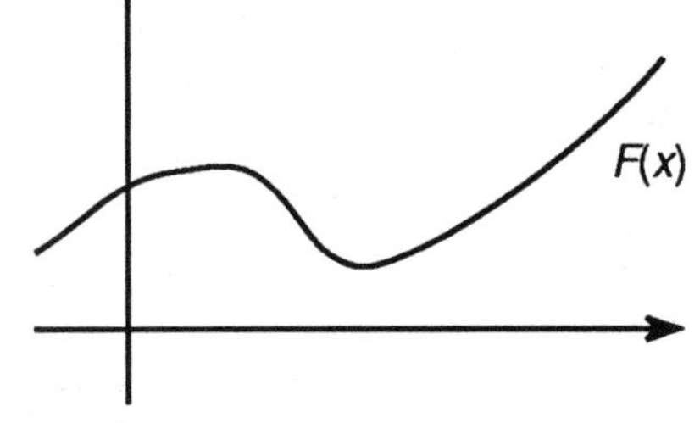

2. Suppose $F' = G' = f$ (That is, F and G are two antiderivatives for f.) If $F(0) = 10, F(1) = 17, G(1) = 12$, what is $G(0)$?

3. If F is an antiderivative for f, and G is an antiderivative for g, give an antiderivative for the following:

(a) $f + g$
(b) $f - g$
(c) kf where k is a constant
(d) $k_1 f + k_2 g$ where k_1 and k_2 are constants
(e) $f + k$ where k is a constant
(f) $f \cdot g$
(g) f/g
(h) $fG + gF$
(i) $\frac{Gf - Fg}{G^2}$
(j) $\frac{Gf - Fg}{F^2}$
(k) $\frac{f}{F}$
(l) $(F)^3 \cdot f$
(m) $f \sin F$
(n) $f \sin F \cos F$
(o) $f(G(x))g(x)$

4. The Fundamental Theorem of Calculus says (in one form) that if f is continuous on an interval I and if $F(x) = \int_a^x f(t)dt$, then $F' = f$. The point $a \epsilon I$ is arbitrary. What happens if a is changed? For a given f the graphs F_1, F_2, F_3 and F_4 result when a is changed.

   (a) Which F_i is obtained when $a = 0$? Why?

   (b) Which F_i is obtained when $a = 1$? Why?

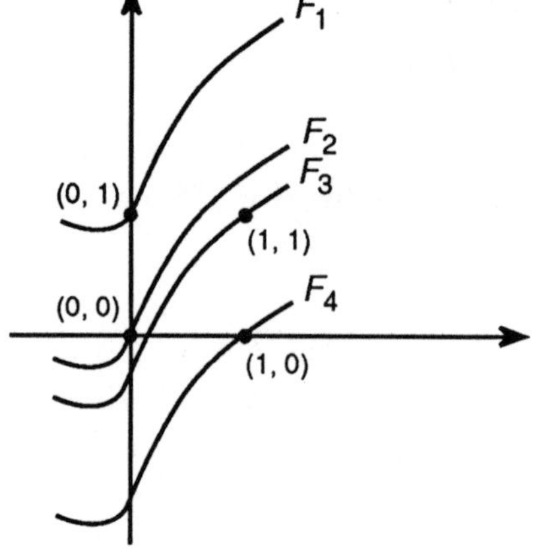

5. Jill says, "I found that $\sin^2\theta$ is an antiderivative for $f(\theta)$." Joe says, "I got $\frac{-1}{2}\cos 2\theta$. I wonder who made a mistake?" It is now your turn to talk. What do you say?

Answers to Antidifferentiation Probes

1. any graph obtained by a vertical shift

2. $F - G$ is constant so $F(1) - G(1) = F(0) - G(0)$ that is, $17 - 12 = 10 - G(0)$, so $G(0) = 5$.

3. (a) $F + G$

   (b) $F - G$

   (c) kF since $(kF)' = kF' = kf$

   (d) $k_1F + k_2G$ since $(k_1F + k_2G)' = k_1F' + k_2G' = k_1f + k_2g$

   (e) $F(x) + kx$ has derivative $F'(x) + k = f(x) + k$

   (f) ? *Not FG*

   (g) ? *Not F/G*

   (h) FG since $(FG)' = F'G + G'F = fG + gF$

   (i) F/G has derivative $\frac{GF'-FG'}{G^2} = \frac{Gf-Fg}{G^2}$

   (j) $-G/F$

   (k) $\ln F$ since $(\ln F)' = \frac{1}{F} \cdot F' = \frac{f}{F}$

   (l) $\frac{F^4}{4}$

   (m) $-\cos F$

   (n) $\frac{1}{2}\sin^2 F($ since $(\frac{1}{2}\sin^2 F)' = 2 \cdot \frac{1}{2}\sin F \cos F \cdot F')$; or similarly, $-\frac{1}{2}\cos^2 F$. (Thus $\frac{1}{2}\sin^2 F - (-\frac{1}{2}\cos^2 F)$ is constant, which we know is $\frac{1}{2}$.)

   (o) $F(G(x))$

4. When a is changed, F changes by a constant. Note that if $F_1(x) = \int_{a_1}^{x} f(t)dt$ then $F_2(x) = \int_{a_2}^{x} f(t)dt = \int_{a_2}^{a_1} f(t)dt + \int_{a_1}^{x} f(t)dt = \int_{a_2}^{a_1} f(t)dt + F_1(x) = k + F_1(x)$ where k is the *fixed* number $\int_{a_2}^{a_1} f(t)dt$.

   (a) F_2. If $F(x) = \int_0^x f(t)dt$, then $F(0) = \int_0^0 f(t)dt = 0$; that is, F passes through the origin.

   (b) F_4, since if $F(x) = \int_1^x f(t)dt$, then $F(1) = 0$.

5. Both are right or both are wrong, since $\sin^2\theta - (-\frac{1}{2}\cos 2\theta) = \sin^2\theta + (\cos^2\theta - \frac{1}{2}) = \frac{1}{2}$, which is constant.

Finding Antiderivatives

Integration Formulas

A. $\int u^n du = \frac{1}{n+1}u^{n+1} + C \quad n \neq -1$ Power Form

B. $\int \frac{du}{u} = \ln \mid u \mid + C$ Logarithm Form

C. $\int e^u du = e^u + C$

D. $\int a^u du = \frac{1}{\ln a}a^u + C$

(C and D: Exponential Forms)

E. $\int \cos u du = \sin u + C$

F. $\int \sin u du = -\cos u + C$

G. $\int \tan u du = -\ln \mid \cos u \mid + C$

(E, F and G: Trigonometric Forms)

Each of the following integrals is easily computed from one of the formulas above when an appropriate factor is inserted in the parentheses. Write in that factor and give the formula that applies from the list.

Example — Formula

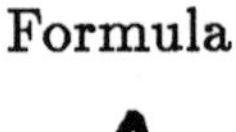

1. $\int (x^2+1)^{\frac{3}{2}}(\ 2x\ )dx$ — A

2. $\int (x^2+1)^{-1}(\qquad)dx$

3. $\int \cos e^{2x}(\qquad)dx$

4. $\int \cos^3 e^{2x}(\qquad)dx$

5. $\int 10^{\ln x}(\qquad)dx$

6. $\int \sqrt{e^{3x+1}}(\qquad)dx$

7. $\int \frac{(\qquad\qquad)}{x^2 + e^x} dx$

8. $\int \frac{(\qquad\qquad)}{(x^2 + e^x)^{100}} dx$

9. $\int (\sin x \cos x^2)^3 (\qquad\qquad\qquad\qquad\qquad) dx$

10. $\int \frac{\sin e^x}{\cos e^x} (\qquad\quad) dx$

11. $\int e^{\cos^2 x} (\qquad\qquad\qquad\qquad) dx$

12. $\int 4^{\cos^2 x \sin x} (\qquad\qquad\qquad\qquad\qquad) dx$

13. $\int \sqrt{x^2 + e^x - \sin x} (\qquad\qquad\qquad\qquad) dx$

14. $\int \frac{1}{\sqrt{x^2 + e^x - \sin x}} (\qquad\qquad\qquad\qquad) dx$

15. $\int \frac{1}{x^2 + e^x - \sin x} (\qquad\qquad\qquad\qquad) dx$

16. $\int \cos(\sin x^2) (\qquad\qquad) dx$

17. $\int \cos^{\frac{3}{2}} (\sin x^2) (\qquad\qquad\qquad\qquad) dx$

18. $\int \frac{1}{\cos(\sin x^2)} (\qquad\qquad\qquad\qquad) dx$

19. $\int (x^3 \sin x)^3 (\qquad\qquad\qquad\qquad) dx$

20. $\int (4x + \sqrt{x})^{-1} (\qquad\qquad) dx$

Answers to Finding Antiderivatives Probes

1. $2x$ $\quad A$

2. $2x$ $\quad B$

3. $2e^{2x}$ $\quad E$ or $-2e^{2x}\sin e^{2x}$ $\quad A$

4. $-2e^{2x}\sin e^{2x}$ $\quad A$

5. $\frac{1}{x}$ $\quad D$ or $10^{\ln x}(\frac{1}{x})$ $\quad A$

6. $3e^{3x+1}$ $\quad A$

7. $2x + e^x$ $\quad B$

8. $2x + e^x$ $\quad A$

9. $\cos x \cos x^2 - 2x \sin x \sin x^2$ $\quad A$

10. $-e^x$ $\quad B$

11. $-2\cos x \sin x$ $\quad C$

12. $\cos^3 x - 2\cos x \sin^2 x$ $\quad D$

13. $2x + e^x - \cos x$ $\quad A$

14. $2x + e^x - \cos x$ $\quad A$

15. $2x + e^x - \cos x$ $\quad B$

16. $2x\cos x^2$ $\quad E$

17. $-\sin(\sin x^2)\cdot 2x\cos x^2$ $\quad A$

18. $-\sin(\sin x^2)\cdot 2x\cos x^2$ $\quad B$

19. $3x^2\sin x + x^3\cos x$ $\quad A$

20. $4 + \frac{1}{2\sqrt{x}}$ $\quad B$

The Definite Integral and Antidifferentiation

The definite integral of f on $[a,b]$, $\int_a^b f(x)dx$, is a number. Geometrically that number can be interpreted as an area. Later we will see how it can be used for various applications–finding volumes, centroids of figures, work done, or the hydrostatic force on a dam face are typical examples. That definite integral was defined as a rather complicated limit, as you have seen.

The Fundamental Theorem of Calculus shows us how to arrive at that limit, at least for a large class of functions, in a much easier way. In particular, assuming f is continuous and supposing we know or can compute a function F for which $F' = f$, then we have $\int_a^b f(x)dx = F(b) - F(a)$. The function F (there is more than one) is called an *antiderivative of f* or the *indefinite integral of f*. So remember, when you are asked to compute a *definite* integral, your answer will be a number. When you are asked to find an *indefinite* integral (or *antiderivative*), your answer will be another function. The Fundamental Theorem shows us how to use indefinite integrals to compute definite integrals.

Finding antiderivatives is simply the process of reversing differentiation. To find the antiderivative of $\sin x$ we ask: "What can I differentiate to get $\sin x$?" The answer is $-\cos x$, or $-\cos x + 10$, or any member of the family of functions $-\cos x + C$ where C is a constant. In fact, that family contains all antiderivatives of $\sin x$, since any two antiderivatives must differ by a constant.

All continuous functions f have antiderivatives (for example, $\int_0^x f(t)dt$), but we may not be able to express them in a way that helps us compute definite integrals. For example, there is no function whose derivative is e^{x^2} that can be written in terms of functions whose value we know. Thus $\int_0^1 e^{x^2}dx = F(1) - F(0)$ is valid but we will not be able to find a useable F. For this reason it is important to remember the definition of the definite integral as a limit, because we use that definition to develop methods on the computer to approximate $\int_0^1 e^{x^2}dx$. These methods are generally called numerical methods of integration, and are used when the Fundamental Theorem doesn't help because we cannot find a useable antiderivative F. Such methods are becoming increasingly important and are covered in full courses called "numerical analysis." One important question is how close the approximations are to the exact value we are seeking.

A final word about the symbol used for integration: The "$\int$" comes from "S" for "sum" from the Riemann or Upper-Lower sum development of the integral. We have already used the "dx" for "differential." Is it a differential here? Later we will see how it can formally be treated as a differential and we get correct results. Until that time it can simply be thought of as part of the symbol we use for the integral. Since $\int_0^1 x^2 dx = \int_0^1 t^2 dt = \int_0^1 q^2 dq = 1/3$, or in general $\int_a^b f(x)dx = \int_a^b f(t)dt$, we call the x, t, or q a "*dummy variable.*" Its name has no effect on the number that we get for the definite integral.

THE FUNDAMENTAL THEOREM

Visualizing the Fundamental Theorem of Calculus

Fundamental Theorem of Calculus:
If f is continuous and F is defined by $F(x) = \int_a^x f(t)dt$, then $F'(x) = f(x)$.

What makes this theorem "fundamental?" Is there a way of visualizing what the theorem says?

You have seen that there are two major tools in calculus, the derivative and the integral. They are both defined by taking certain limits, which on the surface might not appear to be related in any way. After all, why should limits such as

$$\lim_{x \to a} \frac{f(x) - f(a)}{x - a}$$

and

$$\lim_{\|\Delta x\| \to 0} \sum_{i=1}^{n} f(\xi_i)(x_i - x_{i-1})$$

have anything to do with each other?

One reason the Fundamental Theorem of Calculus is fundamental is that it is a *bridge* between differential and integral calculus–it displays the dependence of the two ideas. A second reason is that the theorem is extremely useful. It gives a method of computing definite integrals without having to resort to evaluating limits of the second type shown above. In general, that is very hard to do.

Now let's turn to the second question: Is there a way to visualize the theorem–to intuitively see why the statement is true? Answer the following questions.

Exercises on Visualizing the Fundamental Theorem of Calculus

1. The graph of $y = f(x)$ is shown. Give a geometric interpretation of $F(x+h) - F(x)$ by shading the area it represents. Recall that $F(x) = \int_a^x f(t)dt$, so $F(x+h) = \int_a^{x+h} f(t)dt$.

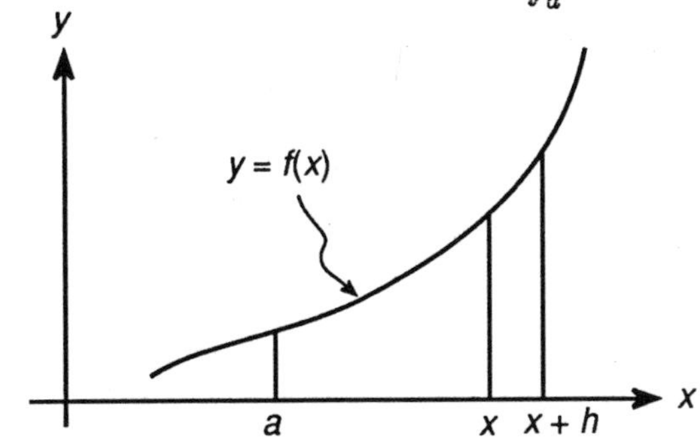

2. Considering your answer to question 1, give a verbal (but geometric) interpretation of the difference quotient $\frac{F(x+h)-F(x)}{h}$.

3. The Fundamental Theorem of Calculus says that $F'(x) = \lim_{h \to 0} \frac{F(x+h)-F(x)}{h} = f(x)$. In light of your answer to question 2, why does this make sense?

Now you *see* the reason for the Fundamental Theorem of Calculus.

Answers to Exercises on Visualizing the Fundamental Theorem of Calculus

1. $F(x+h)$ is the area under the curve from a to $x+h$. From that we subtract $F(x)$, the area under the curve from a to x. Thus $F(x+h) - F(x)$ gives the area of the shaded region.

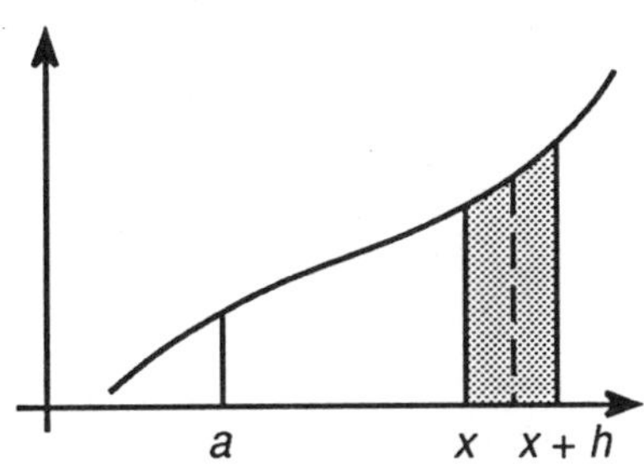

2. This is the area of the shaded region divided by the length of the base of that region. This quotient $\dfrac{F(x+h) - F(x)}{h}$ is the average height of the shaded region; by that we mean the height of a rectangle with the same base and area as the shaded region. It would be a number about the length of the dashed height shown in answer 1, above.

3. As $h \to 0$ the average height of the shaded region, which becomes skinnier, must approach $f(x)$, since the dashed height is always found inside the shaded region. Note the use of the continuity of f in this step. We wish to say the dashed heights approach the height $f(x)$ as the right side of the shaded region moves left; that is, as $h \to 0$. Continuity ensures that this happens.

APPLICATIONS OF THE INTEGRAL

Work, Hydrostatic Force, and Centroids

A. The work done in emptying a tank of liquid through a pipe out the top of the tank is the same as the work done in raising all the liquid from the centroid of the tank to the top of the pipe; that is, as far as work is concerned we may assume all the liquid is located at the depth where the centroid lies.

B. The hydrostatic force on a vertical plate in a liquid is the same as the force on a plate of equal area whose total area is at the depth of the centroid; that is, as far as hydrostatic force is concerned, we may assume the entire plate is at the depth where the centroid lies.

These two statements are both interesting and useful; they help us to have an intuitive feel for problems of work and force. Let's first review how we justify the two statements.

Justification of Statement A. Suppose the tank in the figure is full of liquid of density ρ, has cross section area $A(x), 0 \leq x \leq D$, and the pipe at the top has length $\ell \geq 0$.

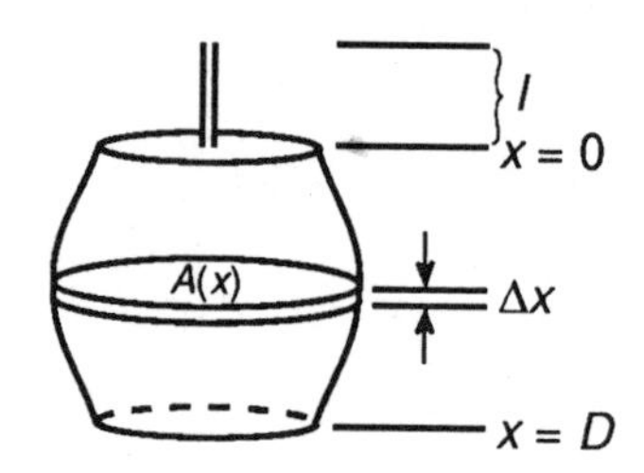

$$\begin{aligned}
\text{We see } A(x)\Delta x &\approx \text{volume of a } \Delta x \text{ slice of liquid at depth } x \\
\rho A(x)\Delta x &\approx \text{weight of that slice of liquid} \\
(\ell + x)\rho A(x)\Delta x &\approx \text{work required to raise that slice to the top of the pipe} \\
(*) \int_0^D (\ell + x)\rho A(x)dx &= \text{work required to empty the tank through the pipe}
\end{aligned}$$

Now, to find the centroid of the liquid we compute

$$\begin{aligned}
x\rho A(x)\Delta x &\approx \text{moment of the slice at depth } x \text{ about the plane } x = 0 \\
\int_0^D x\rho A(x)dx &= \text{total moment of liquid about } x = 0 \\
\int_0^D \rho A(x)dx &= \text{total weight of liquid}
\end{aligned}$$

Thus $\bar{x} = \dfrac{\rho \int_0^D xA(x)dx}{\rho \int_0^D A(x)dx}$ = depth of centroid (Note the ρ's cancel; that is, the centroid for water is the same as the centroid for mercury and ρ is beside the point in our computation of centroid.)

Now the work done to raise all the liquid from depth $\bar{x}$ would be

$$(\bar{x} + \ell)(\rho \int_0^D A(x)dx) = \left(\frac{\rho \int_0^D xA(x)dx}{\rho \int_0^D A(x)dx} + \ell\right)(\rho \int_0^D A(x)dx)$$

$$= \rho \int_0^D xA(x)dx + \ell\rho \int_0^D A(x)dx = \rho \int_0^D (x + \ell)A(x)dx$$

But this agrees with $(*)$ and we have justified statement A.

Fill in the following blanks for a justification of statement B. Assume the vertical plate shown has cross section width $w(x)$ at depth $x, a \leq x \leq b$, and the density of the liquid is ρ.

________ $\approx$ area of slice of plate at depth x

________ $\approx$ pressure on that slice

________ $=$ total area of the plate

(*) ________ $=$ total force on the plate

x = 0

x = a

w(x)

x = b

To compute centroid depth we see:

________ $\approx$ moment of the slice of area of plate at depth x about $x = 0$ (assume the plate's density is 1 unit per square unit of area)

________ $=$ total moment of plate about $x = 0$

________ $=$ $\bar{x}$

(**) ________ $=$ force on the plate if all the plate were at depth $\bar{x}$

Now show (**) can be reduced to (*)

Justification of Statement B

$$w(x)\Delta x \approx \text{area of the slice of plate at depth } x$$

$$\rho x w(x)\Delta x \approx \text{force on that slice}$$

$$\int_a^b w(x)dx = \text{total area of the plate}$$

$$(*) \int_a^b \rho x w(x)dx = \text{total force on the plate}$$

$$x w(x)\Delta x \approx \text{moment of the slice of area of width } \triangle x \text{ of plate at depth } x \text{ about } x = 0$$

$$\int_a^b x w(x)dx = \text{total moment of plate about } x = 0$$

$$\bar{x} = \frac{\int_a^b x w(x)dx}{\int_a^b w(x)dx}$$

$$(**) \rho\bar{x}\int_a^b w(x)dx = \rho\left(\frac{\int_a^b x w(x)dx}{\int_a^b w(x)dx}\right)\int_a^b w(x)dx = \rho\int_a^b x w(x)dx \text{ which is } (*).$$

Exercises on Work, Hydrostatic Force, and Centroids

1. In (a), (b), and (c) each, below, three tanks are shown. List them in order of least to most work required to empty the tanks. The ends of the tanks all have equal area.

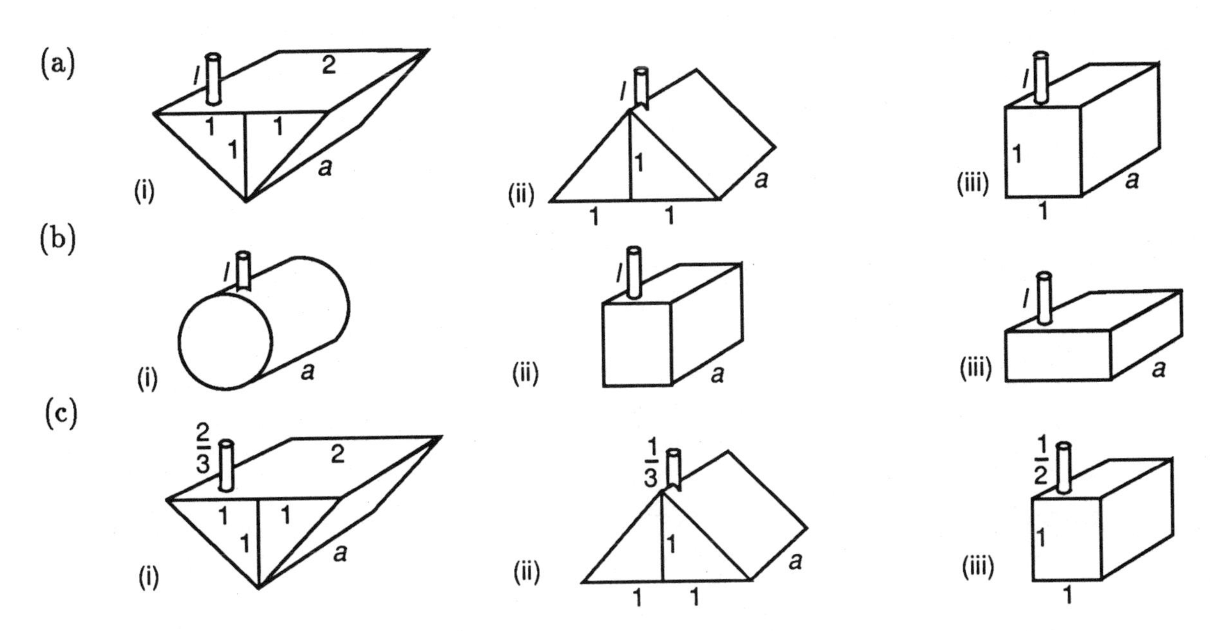

2. In each of the figures, (a), (b), and (c) below, order the plates from the one with least hydrostatic force to the one with greatest force. The shaded areas are all equal.

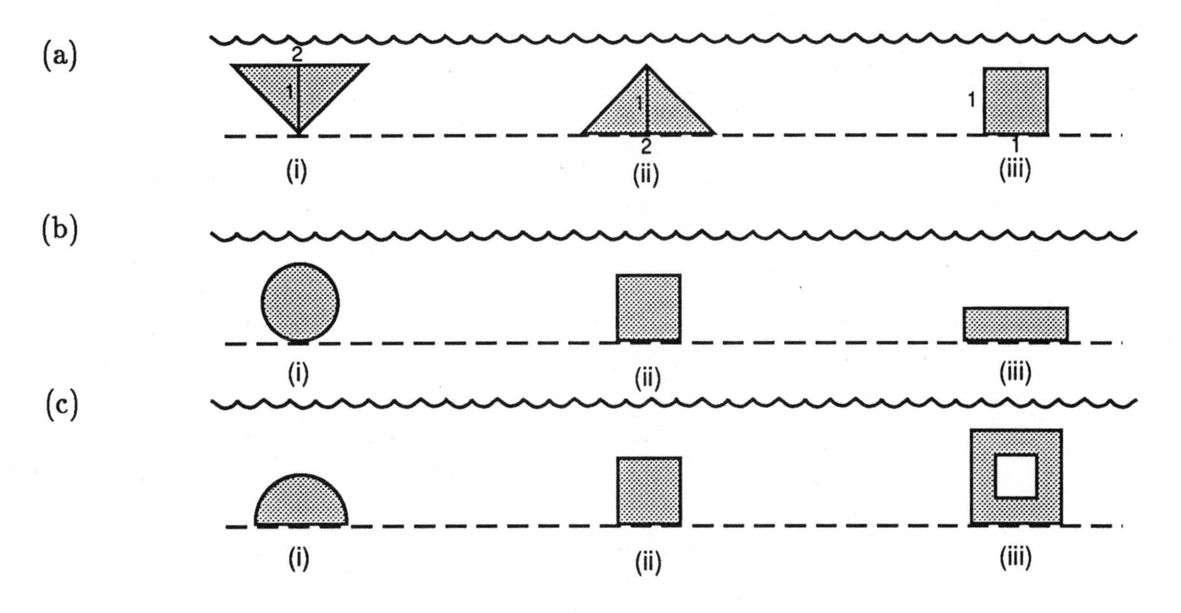

3. Suppose a vertical plate has its top at the level of the liquid in a tank. The bottom of the plate is at depth D. If the plate is raised until the bottom of the plate is at depth $D/2$, prove that the force in the second position is less than or equal to half the original force regardless of the shape of the plate.

Answers to Exercises on Work, Hydrostatic Force, and Centroids

1. (a) (i), (iii), (ii) from least to most (just compare location of centroids)

   (b) Assume the square end is one by one, so the $\bar{x}$ there is $1/2$. $\bar{x}$ in (iii) is smaller than $\frac{1}{2}$, for the circle $\bar{x}$ is just the radius of a circle whose area is 1 unit; that is, $1 = \pi(\bar{x})^2$. So $\bar{x} = \frac{1}{\sqrt{\pi}} > \frac{1}{2}$. So in order of least to most work to empty, we have (iii), (ii), (i).

   (c) They are all equal.

2. (a) (i) (iii) (ii)

   (b) (i) (ii) (iii)

   (c) (iii) (ii) (i)

3. Think of the plate in two pieces, that with area A_1 whose original depth lies between 0 and $D/2$, and that with area A_2 whose depth lies between $D/2$ and D. Let the centroid of A_1 originally lie at depth x_1 and the centroid of A_2 at depth x_2. Then the original pressure is $\rho(A_1x_1 + A_2x_2)$. The other pressure is $\rho A_2(x_2 - \frac{1}{2}D)$.

Now since $x_2 \leq D$ we have $2x_2 - D \leq x_2$. Therefore,

$$\rho A_2(x_2 - \frac{1}{2}D) = \frac{1}{2}\rho A_2(2x_2 - D) \leq \frac{1}{2}\rho A_2x_2 \leq \frac{1}{2}\rho(A_1x_1 + A_2x_2), \text{as required.}$$

Cauchy's and Barbier's Formulas

Augustin Louis Cauchy (1789-1857) showed that an interesting and beautiful connection exists between the length of a convex curve and its average width. When applied to curves of constant width, it reduces to Barbier's formula. These formulas illustrate how calculus can be a tool for geometric discovery.

Some terms will be useful in stating the results. A line through a point P on a plane curve C is a *support line* if the curve lies to just one side of the line. If every point of a curve has one (or more) support lines, the curve is *convex*. The distance between two parallel supporting lines is called the *width* of the curve in the direction perpendicular to the lines. If the width is the same in all directions, the curve is said to be of *constant width*.

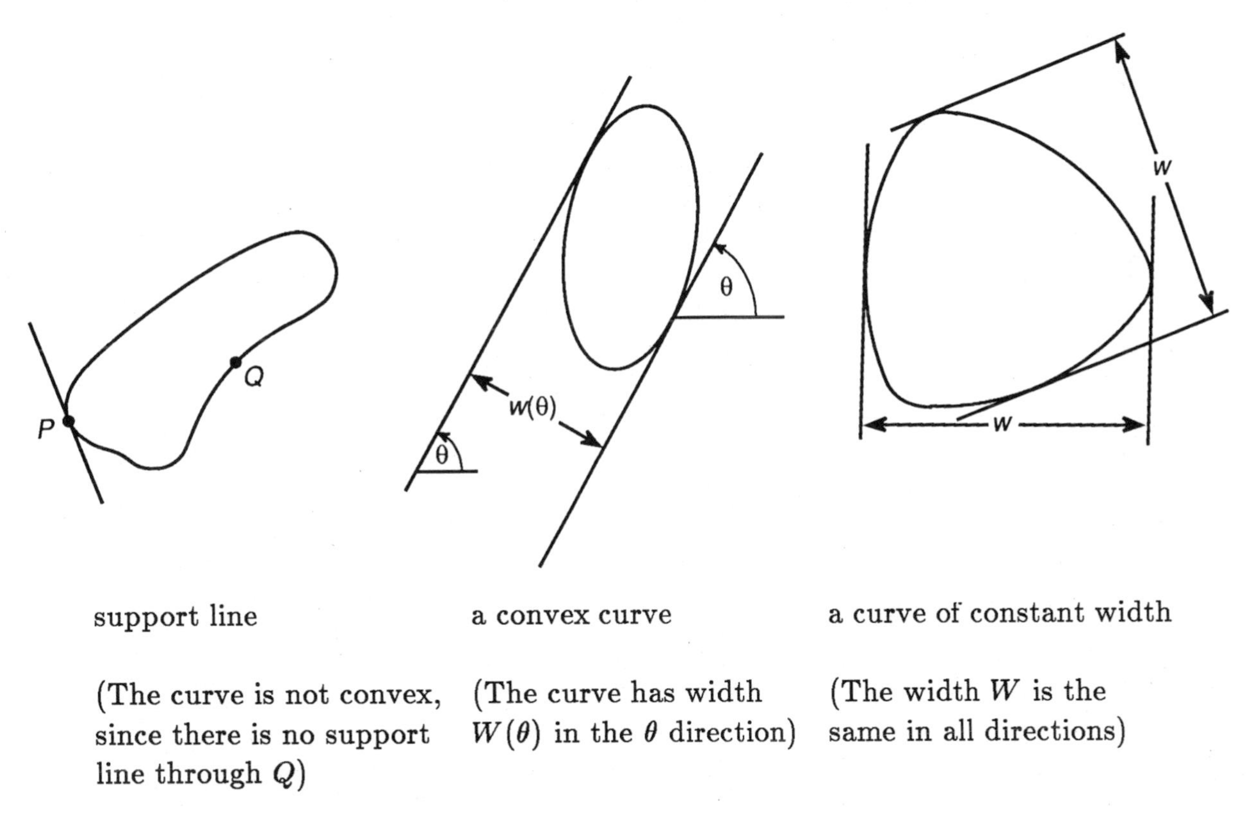

support line — (The curve is not convex, since there is no support line through Q)

a convex curve — (The curve has width $W(\theta)$ in the θ direction)

a curve of constant width — (The width W is the same in all directions)

Notice that $W(\theta)$ can be viewed as the length of the curve's shadow cast by parallel rays of light. Since C can be approximated by a polygon, we first investigate the average shadow length cast by a line segment in all directions. The general result then follows by the standard limiting process basic to calculus.

1. Suppose a line segment has length ℓ. What is its average shadow length where the average is taken over all directions?

Answer

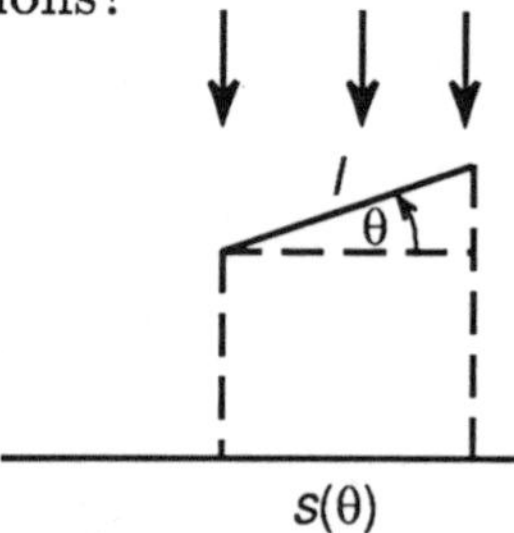

As the figure shows, the segment $\overline{OA}$ of length ℓ casts a shadow of length $\ell \mid \cos\theta \mid = s(\theta)$. Thus the average length $\overline{s}$ of its shadow is given by

$$\overline{s} = \frac{1}{2\pi}\int_0^{2\pi} \ell \mid \cos\theta \mid d\theta = \frac{4\ell}{2\pi}\int_0^{\pi/2} \cos\theta d\theta = \frac{2\ell}{\pi}.$$

2. What is the average width $\overline{W}$ of a convex polygon?

Answer

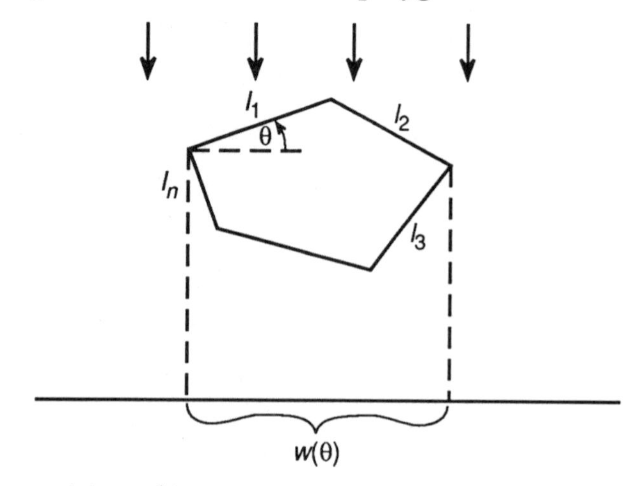

Here we see that $W(\theta) = \frac{1}{2}[s_1(\theta) + \cdots + s_n(\theta)]$. The reason for the $\frac{1}{2}$ factor on the right is that the shadow is doubly covered by the individual shadows from two segments. From part 1 we see that

$$\overline{W} = \frac{1}{2}[\overline{s}_1 + \overline{s}_2 + \cdots + \overline{s}_n] = \frac{1}{2}[\frac{2\ell_1}{\pi} + \frac{2\ell_2}{\pi} + \cdots + \frac{2\ell_n}{\pi}]$$

Since $L = \ell_1 + \ell_2 + \cdots + \ell_n$ is the length of the polygon, we get

$$\overline{W} = \frac{L}{\pi}.$$

Solving for L we see that $L = \pi\overline{W}$. That is, the length of the convex polygon is π times its average width. Since any convex curve may be approximated by a convex polygon, we deduce *Cauchy's Formula:* Let $W(\theta), 0 \leq \theta \leq 2\pi$ be the width of a convex curve in direction θ. Then the length L of the curve is given by

$$L = \pi\overline{W} = \int_0^{\pi} W(\theta)d(\theta).$$

If a convex curve has constant width W, then $\overline{W} = W$ and so the length of the curve is πW. This is:

Barbier's Formula: A convex curve of constant width W has length $L = \pi W$.

3. Verify Barbier's Theorem for the Reauleaux* Triangles shown below, each of which is formed by circular arcs centered at the vertices of an equilateral triangle.

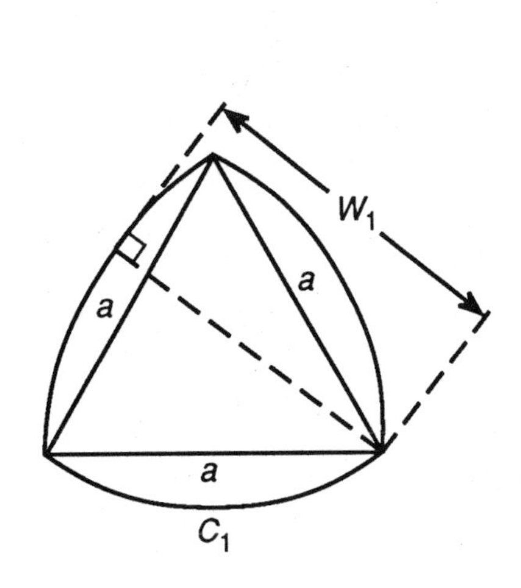

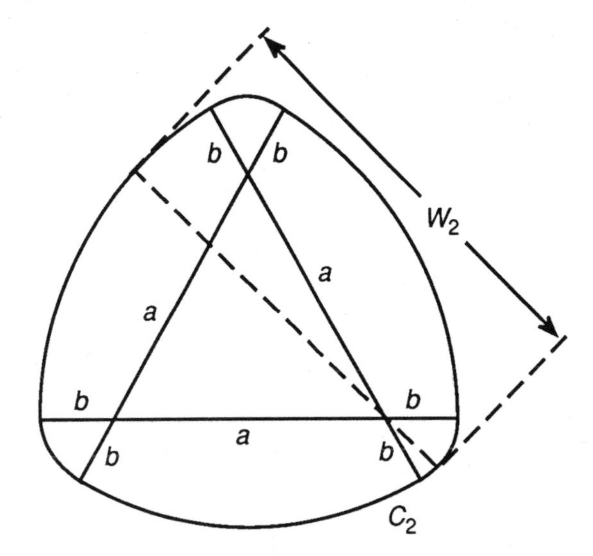

Answer It is clear from the figures that the constant widths of C_1 and C_2 are respectively $W_1 = a$ and $W_2 = a + 2b$. A circular arc of angle 60° and radius R has length $\frac{1}{6}(2\pi R) = \frac{1}{3}\pi R$, and so

$$\begin{aligned} L_1 &= \text{length } C_1 = 3(\tfrac{1}{3}\pi a) = \pi a = \pi W_1 \\ L_2 &= \text{length } C_2 = 3[\tfrac{1}{3}\pi(a+b) + \tfrac{1}{3}\pi b] = \pi(a+b+b) = \pi W_2 \end{aligned}$$

*Reauleaux was a French engineer who designed some 50 rotary engines. Strangely enough, none incorporated the Reauleaux triangle, which is the shape of the rotor in the successful rotary engine of Felix Wankel.

Cavalieri's Principle

Bonaventura Cavalieri (1598-1647) was a student and associate of Galileo. Galileo in fact suggested to Cavalieri that he investigate the use of "indivisibles"* for the determination of areas and volumes. This investigation eventually resulted in the book *Geometria Indivisibilibus Continuorum*, which in many ways is the first textbook on what we now call methods of integration.

Cavalieri's Principle for Areas. A planar surface, in Cavalieri's way of thinking, consists of an indefinite number of uniformly-spaced parallel line segments–these are the so-called "indivisibles." Thus the area of two surfaces can be compared by comparing corresponding line segments in each figure. In particular, if each line segment of one figure is matched by a line segment of the same length in a second figure, then the two figures have the same area.

In modern notation Cavalieri's principle is equivalent to the result shown in the figure below.

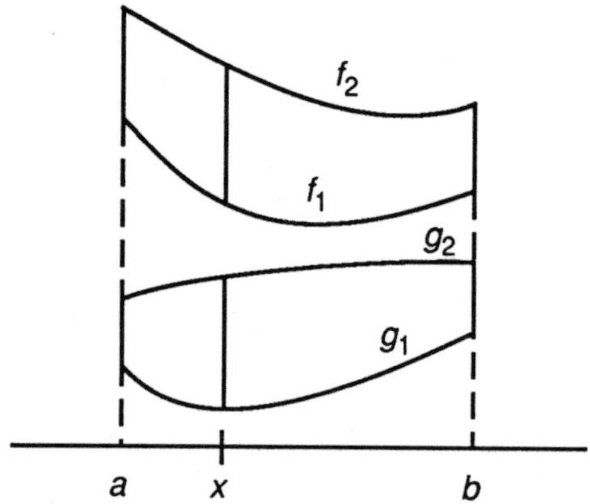

Here one figure lies between the graphs of two continuous functions f_1, f_2 defined on $a \leq x \leq b$, and a second figure is similarly defined by the functions g_1 and g_2. If every vertical line intersects line segments of equal length in both regions, then the two figures have equal area. In symbols, the principle is simply this:

$$\text{If } g_2(x) - g_1(x) = f_2(x) - f_1(x) \text{ for } a \leq x \leq b\,,$$

$$\text{then } \int_a^b (g_2(x) - g_1(x))\, dx = \int_a^b (f_2(x) - f_1(x))\, dx.$$

* "Indivisibles" or "infinitesimals" was an ill-defined and misleading concept until quite recently, when an elaborate set-theoretic structure firmed up what are called the hyperreal numbers. The idea of calculus based on hyperreal numbers is called *nonstandard analysis*.

Example. The area of a cycloidal arch.

Recall that a cycloid is the curve generated by a point on the rim of a wheel that rolls without slipping on a straight line, as shown by the heavy line in the figure below.

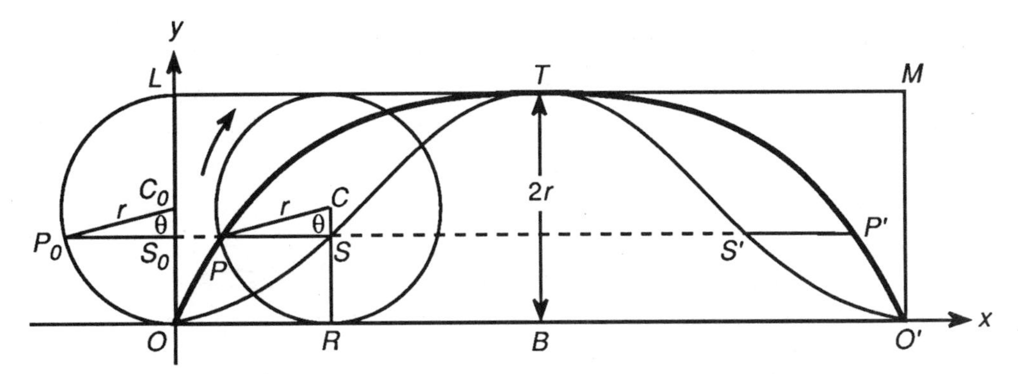

The assumption that the wheel doesn't slip tells us that OR is the length of the arc $\overset{\frown}{PR}$, and so

$$OR = r\theta.$$

It is now easy to find the coordinates of the points P on the cycloid, and the point S as shown in the figure, namely

$$\begin{aligned} P &= (r\theta - r\sin\theta, r - r\cos\theta) \\ S &= (r\theta, r - r\cos\theta). \end{aligned}$$

The locus of points S is a sine wave, and it is of historic interest to note that this is the first appearance, in the early 17th century, of the sine wave. For many years, the sine wave was called "the companion curve to the cycloid."

It is geometrically clear from symmetry that the area beneath the sine wave $OSTS'O'$ is half that of the rectangle $OO'ML$. Since the rectangle has width $OO' = 2\pi r$ and height $OL = 2r$, its area is $4\pi r^2$ and so we conclude that

$$\text{area under companion curve} = 2\pi r^2.$$

Now we use Cavalieri's Principle. Since $P_\circ S_\circ = PS = P'S' = r\sin\theta$ at each horizontal level, each of the two shaded regions between the cycloid and the companion curve has the area of the half-circle. Thus we have:

$$\text{area between companion curve and cycloid} = \pi r^2.$$

The total area beneath the cycloidal curve is now evident:

$$\text{area under cycloid} = 3\pi r^2.$$

Since the area of the circle is πr^2, we see that the area under a cycloid is precisely three times the area of the circle that was used to generate it. The proof above was given by Gilles Roberval (1602-1675) in 1634. Roberval independently introduced the Cavalierian Principle of "indivisibles."

Cavalieri's Principle for Volumes. For Cavalieri, a solid is viewed as consisting of infinitely many uniformly-spaced planar regions. If two solids (as shown below) have corresponding planar regions of equal area, $A_1(x) = A_2(x)$, then Cavalieri asserts that the volumes of the solids must also be equal.

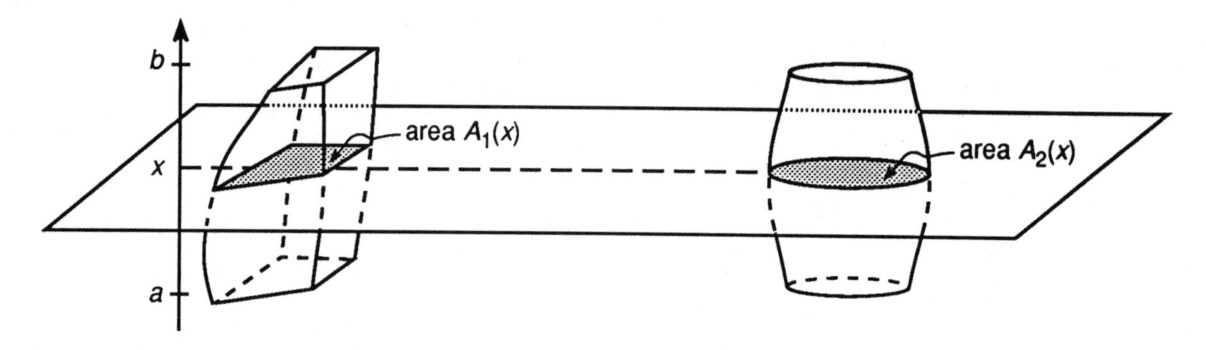

In modern terms this is justified by dividing the solid into parallel slabs and forming a Riemann sum of the form

$$V_1 \approx \sum_{i=1}^{n} A_1(x_i)\Delta x_i.$$

Taking the limit in the usual way, the exact volume is given by the integral

$$V_1 = \int_a^b A_1(x)dx.$$

It follows easily that if $A_1(x) = A_2(x)$ for $a \leq x \leq b$, then the volumes are also equal, $V_1 = V_2$.

Example. The volumes of cored spheres.

Consider this problem: A solid sphere has a cylindrical hole drilled symmetrically through it, leaving a "pitted olive" of height $2h$. What is the volume of the cored sphere?

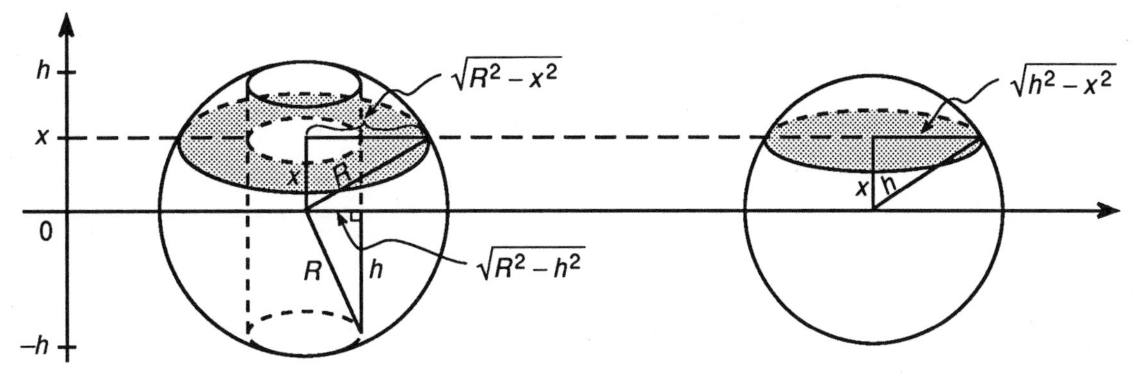

The problem seems to lack sufficient information for its solution, since the radius of the sphere is not given. As the figure above shows, in the extreme case that the diameter of the sphere is $2h$, there would in fact be no core at all, and so the volume is that of a full sphere, namely $\frac{4}{3}\pi h^3$.

To see that the cored sphere of height $2h$ and of unknown radius R has the same volume as the solid sphere of radius h, we must compute the shaded area $A_1(x)$ between the concentric circles. By the Pythagorean theorem, the inner and outer radii are $\sqrt{R^2 - h^2}$ and $\sqrt{R^2 - x^2}$. Thus we get

$$A_1(x) = \pi(\sqrt{R^2 - x^2})^2 - \pi(\sqrt{R^2 - h^2})^2 = \pi(h^2 - x^2).$$

The area of the corresponding circle in the solid sphere is

$$A_2(x) = \pi(\sqrt{h^2 - x^2})^2 = \pi(h^2 - x^2), -h < x < h.$$

Thus $A_1(x) = A_2(x)$ and so by Cavalieri's Principle for volumes we conclude that

$$V_1 = V_2 = \frac{4}{3}\pi h^3.$$

It is hard to believe that cored spheres of equal height $2h$ are the same volume whether they are drilled through an orange or the moon!

Exercises on Cavalieri's Principle

1. A right circular cylinder of radius R and height $2R$ will just contain the sphere of radius R. Suppose the cylinder is "coned out" by removing the two cones of radius R and height R as shown below. Show that the coned-out cylinder and the sphere have the same volume.*

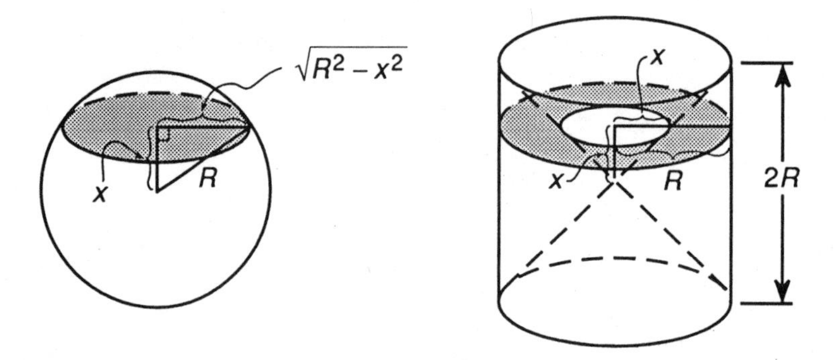

2. Suppose you know (as Archimedes did) that the volume of a cone is one-third the volume of the cylinder that just contains it. Use this to show that the sphere occupies two-thirds of the volume of the right circular cylinder that just contains it.

3. Use the result of problem 2 to obtain the formula for the volume of a sphere, assuming you already know the formula *base area* $\times$ *height* for the volume of a cylinder.

4. The area of a sphere is $4\pi R^2$. Show that the area ratio of the sphere to the enclosing cylinder is also 2 : 3, another discovery by Archimedes.

5. Use Cavalieri's Principle to compare the volumes of the two regions below. On the left is one octant of the region common to two right circular cylinders of radius r whose axes intersect at right angles. On the right is a cube of side r from which a square-based pyramid has been removed.

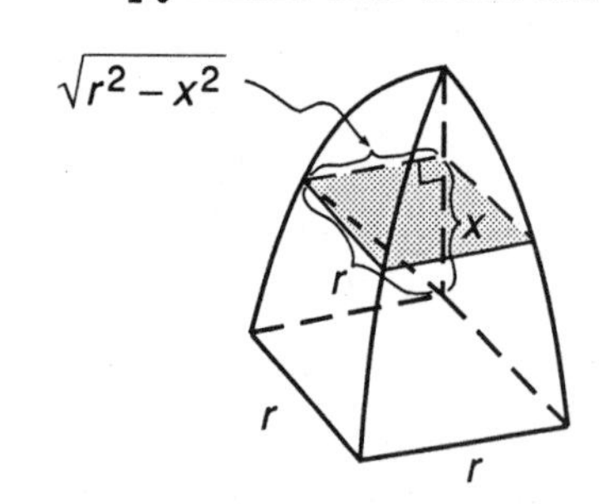

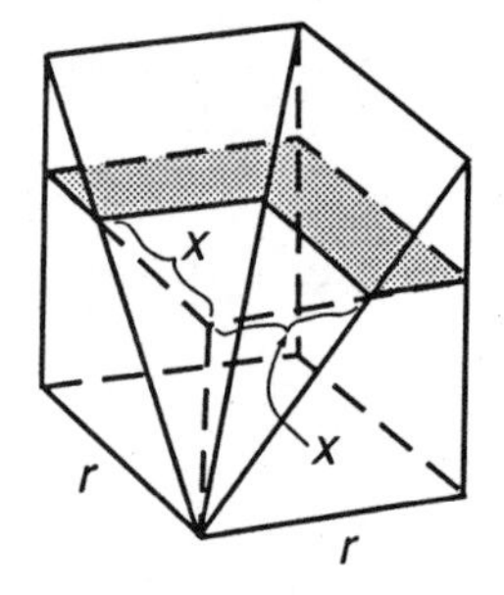

---

* *Historical Remark.* Archimedes viewed the work in his book *On the Sphere and the Cylinder* with justifiable pride. He asked that his tombstone be engraved with a sphere and its circumscribing cylinder, and Marcellus carried out the wish. For many years the monument was lost, but it was found and restored by Cicero in 75 B.C., only to be lost again. Amazingly, the tomb was once again rediscovered, during the excavation in 1965 for the foundation of a hotel in Syracuse, southwest Sicily.

Answers to Exercises on Cavalieri's Principle

1. From the left-hand figure and use of the Pythagorean Theorem, the circular section at level x in the sphere has area $A_1(x) = \pi(\sqrt{R^2 - x^2})^2 = \pi(R^2 - x^2)$. The annular region in the coned-out cylinder has inner and outer radii of x and R, and so its area is $A_2(x) = \pi R^2 - \pi x^2 = A_1(x)$. By Cavalieri's Principle, both solids have the same volume.

2. Since a cone has one-third the volume of the corresponding cylinder, the coned-out cylinder has two-thirds the volume of the full cylinder. From the result of problem 1, the volume of the sphere is therefore exactly two thirds of its circumscribing cylinder.

3. The cylinder that circumscribes a sphere of radius R has base area πR^2 and height $2R$, giving it a volume of $(\pi R^2)2R = 2\pi R^3$. The volume of the sphere, from problem 2, is two thirds of this, giving $\frac{2}{3}(2\pi R^3) = \frac{4}{3}\pi R^3$.

4. The area of the cylinder that circumscribes a sphere of radius R is given as the area of the circular bases (each a circle of radius R) and the lateral area (which can be "unrolled" to be a rectangle of height $2R$ and length $2\pi R$). The area of the cylinder is therefore
$$\pi R^2 + \pi R^2 + (2R)(2\pi R) = 6\pi R^2.$$
The ratio of the area of the sphere to the area of its circumscribing cylinder is then $4\pi R^2 : 6\pi R^2 = 2 : 3$.

5. In the left-hand figure at level x, the region is a square of side $\sqrt{r^2 - x^2}$, and so the area of the square is $A_1(x) = r^2 - x^2$. The corresponding L-shaped region on the right is an r by r square with an x by x square removed, and so the angular region has area $A_2(x) = r^2 - x^2 = A_1(x)$. Thus both regions have the same volume, V. The volume of the right-hand solid is easy to compute: three congruent pyramids form the complete cube, and since one pyramid has been removed, we see $V = \frac{2}{3}r^3$. It's surprising to realize that the volume on the left has been found *without integration!* (You have probably seen this problem in the exercises in your text.) Notice that the region on the left occupies $\frac{2}{3}$ of the volume of the circumscribing cube–Archimedes would like that!

The Big Picture

Sequences

By now you have seen two important limits:

$$f'(x) = \lim_{h \to 0} \frac{f(x+h) - f(x)}{h} \text{ for the derivative, and}$$

$$\int_a^b f(x)dx = \lim_{\|\Delta x\| \to 0} \sum_{i=1}^{n} f(\varsigma_i)\Delta x_i \text{ for the integral.}$$

We are now ready for a third limit, the limit of sequences. If $a_1, a_2, a_3, \cdots$ is the sequence $1, 1/2, 1/3, 1/4, \cdots, 1/n, \cdots$ would you agree that $\lim_{n \to \infty} a_n = 0$? Let's make it rigorous.

Let $a_1, a_2, a_3, \cdots$ be a sequence of real numbers. Then $\lim_{n \to 0} a_n = L$ if and only if for every $\epsilon > 0$, there is an N so that if $n > N$ then $\mid a_n - L \mid < \epsilon$.

Your experience with ϵ, δ limits will help. The ϵ here plays the same role it did before: ϵ is a "tolerance" for how far terms are permitted to stray from the limit L. But not all terms $a_1, a_2 \cdots$ must be within ϵ of L, only those in a "tail" $a_{N+1}, a_{N+2}, \cdots$ must be. The N here plays the same role δ did in continuity arguments. It restricts when the inequality $|a_n - L| < \epsilon$ must hold, namely for $n > N$.

Recall also, $\epsilon > 0$ came first and we had to accept any $\epsilon > 0$. Then once ϵ was known, we determined an appropriate δ. The game is the same here. First comes an arbitrary $\epsilon > 0$. We now find a suitable N for that ϵ. The role of N can be illustrated as follows. Someone has given us a challenge $\epsilon > 0$. The definition promises an appropriate N for every $\epsilon > 0$, so we have to produce an N for the given ϵ. The choice of N provides a "cutoff point" in the sequence in the following way:

$a_1, a_2, \cdots, a_N$	$a_{N+1}, \cdots,$
forget these terms $\leftarrow$	$\rightarrow$ from here out we promise *all* terms are within ϵ of L. That is, when $n > N$, we promise $\mid a_n - L \mid < \epsilon$.

To go back to the sequence $1, 1/2, 1/3, 1/4, \cdots$ where $L = 0$, suppose $\epsilon = .01 = \frac{1}{100}$. Then let's choose $N = 100$ so

$1, \frac{1}{2}, \frac{1}{3}, \cdots, \frac{1}{100},$	$\underbrace{\frac{1}{101}, \frac{1}{102}, \frac{1}{103}, \cdots}$
forget these a_{100}	all of these are within $\frac{1}{100}$ of 0; that is, $\mid a_n - 0 \mid < .01$ when $n > N = 100$.

Question: For $\epsilon = .01$ could we have chosen $N = 101$? (Yes) What other Ns would work for $\epsilon = .01$? What N could be chosen for $\epsilon = .01234$? (Wouldn't $N = 100$ work for that ϵ too?)

Let's see how the definition can be used to show that the limit of $1, \frac{1}{2}, \frac{1}{3}, \frac{1}{4}, \cdots$ cannot be -1. If $\epsilon = 1.5$, then we can choose $N = 2$. (Think of all of this on the number line.)

$$1, \tfrac{1}{2}, \;\Big|\; \underbrace{\tfrac{1}{3}, \tfrac{1}{4}, \tfrac{1}{5}, \cdots}$$

a_2 | all of these are within 1.5 of -1, that is, $| a_n - (-1) | = | a_n + 1 | < 1.5$ for $n > N$.

For $\epsilon = 1.5$ we have found an N. But the definition requires an N to be found for *any* $\epsilon > 0$, not just one ϵ. What if we are given $\epsilon = .5$? Is there any cutoff beyond which all terms are within .5 of the "limit" $L = -1$? They would have to be between -1.5 and $-.5$ for that to happen. Since all $a_n > 0$, there is no N for $\epsilon = .5$. The definition requires an N for every ϵ, not just some of them, and so $\lim_{n\to\infty} a_n$ is not -1.

Qualitative statements like the following should make sense. (Think of "cutoff points".)

> "In general as ϵ gets smaller, N will get bigger."
> "If N works for one ϵ, it also works for any larger ϵ."
> "If N works for an ϵ, any bigger N works for the same ϵ."

Do they make sense? If so, you are getting a feel for the ϵ, N game.

One other thing needs to be previewed.

The sequence $a_1, a_2, \cdots a_n, \cdots$ is a Cauchy sequence if for every $\epsilon > 0$ there is an N so that if $n > N$ and $m > N$, then $| a_n - a_m | < \epsilon$.

Cauchy sequences (named after a 19th century mathematician) are important because they are precisely the convergent sequences (this has been proved). But since we already have a definition for convergent sequences, why do we need a second rendition for a convergent sequence? Here is the important answer. In practice, finding the limit L can be very difficult and we may only want to be sure that the sequence converges, not really caring what the limit L is. This situation often arises. Note that in the original definition we had to look at $| a_n - L |$. Terms were compared to L so we had to know what L was. In the definition of a Cauchy sequence L never appears. We examine $| a_n - a_m |$; that is, we compare terms with each other and can show a sequence is Cauchy, and therefore convergent, even without knowing L. To illustrate Cauchy sequences, suppose $\epsilon > 0$ has been given to us; we had no say in its choice. We have to find a suitable N for that $\epsilon > 0$.

$a_1, a_2, \cdots a_N,$	$\underbrace{a_{N+1}, a_{N+2}, a_n, \cdots, a_m \cdots}$
forget these	any two of these are within ϵ of each other; that is, if $n > N$ and $m > N$, then $\mid a_n - a_m \mid < \epsilon$.

Sequences

Recall the definition: A sequence $\{a_n\}$ converges to L if and only if for every $\epsilon > 0$ there is an N so that if $n > N$ then $| a_n - L | < \epsilon$.

1. Give an example of a sequence with no converging subsequence.

2. Give an example of a sequence such that every subsequence converges.

3. Give an example of a sequence for which there are exactly *two* limits to which subsequences converge.

4. Give an example of a sequence for which there are infinitely many limits to which subsequences converge.

5. If, for $\epsilon = 1$, N can be chosen to be 487, then for which other ϵ can N be chosen to be 487?

6. If, for $\epsilon = 1$, N can be chosen to be 487, for $\epsilon = 1$ what other N could be chosen?

7. Suppose for *any* $\epsilon > 0$, N can be chosen to be 410 but not 409. What can you say about the sequence?

8. True-False. If *every* subsequence of $\{a_n\}$ converges to some limit, those limits must all be the same and the original sequence converges to that common limit.

9. True-False. If one subsequence of $\{a_n\}$ converges to a, another to b, and $a \neq b$, then the sequence $\{a_n\}$ *must* diverge.

10. Consider the sequence 1,2,1,2,1,2,... (which we know diverges). If we are under the mistaken impression that the sequence converges to $L = 1.5$ and let $\epsilon = \frac{2}{3}$, can we find a corresponding N? What does that imply about the statement $\lim_{n \to \infty} \{a_n\} = 1.5$?

11. How can we show that 1,2,1,2,1,2,... does not converge to 1.5? For which ϵ can no N be found?

12. In general as ϵ gets smaller, N gets ____________.

13. If the test for convergence fails for some given $\epsilon_0 > 0$ (no N works), then the test fails for which other ϵ?

14. True-False. If $\{a_n\}$ and $\{b_n\}$ both diverge, then $\{a_n + b_n\}$ must diverge.

15. True-False. If $\{a_n\}$ converges and $\{b_n\}$ diverges, then $\{a_n + b_n\}$ must diverge.

Answers to Sequences Probes

1. 1,2,3,4,...
2. Any converging sequence by one of our theorems
3. 1,2,1,2,1,2,...
4. 1,2,1,2,3,1,2,3,4,1,2,3,4,5,...has subsequences converging to each natural number.
5. All $\epsilon \geq 1$, so if the test of the definition is passed for $\epsilon = 1$, it is also passed for all $\epsilon \geq 1$.
6. Any $N > 487$
7. $a_{410} \neq a_{411}$ but $a_{411} = a_{412} = a_{413} = \cdots$
8. True
9. True, for $\epsilon = \left| \frac{a-b}{2} \right|$ no N can be found.
10. Yes, $N = 1$ will work because $| a_n - 1.5 | = \frac{1}{2} > \frac{2}{3}$ for every n. All that says is that the limit *may be* 1.5; not all ϵ have been examined. (See 11.)
11. Any $\epsilon \leq \frac{1}{2}$
12. Larger
13. Any $\epsilon \leq \epsilon_0$. If you can't stay within ϵ_0 of the limit, you can't stay within any smaller ϵ of that limit.
14. False. Let $\{a_n\}$ be 0,1,0,1,0,1,...and $\{b_n\}$ be 1,0,1,0 .... Then $\{a_n + b_n\}$ is the sequence 1,1,1,1,...
15. True. If $\{a_n+b_n\}$ converges and $\{a_n\}$ converges, then $\{(a_n+b_n)-a_n\}$ must converge (the difference of converging sequences converges), that is, $\{b_n\}$ converges.

Nicholas of Cusa's Method of Evaluating π

With the obvious exception of numbers like 0, 1, $\frac{1}{2}$, the most interesting and useful constant is surely π, defined as the ratio of the circumference of a circle to its diameter. Early approximations of π can be found in the Rhind Mathematical Papyrus (c.1650 B.C.) and Archimedes's book *Measurement of a Circle.* In modern notation the Rhind Papyrus gives π as $4(\frac{8}{9})^2$, while Archimedes showed that $3\frac{10}{71} < \pi < 3\frac{1}{7}$. Nicholas of Cusa (1401-1464) invented a sequential method of obtaining approximations to π, which we now describe.

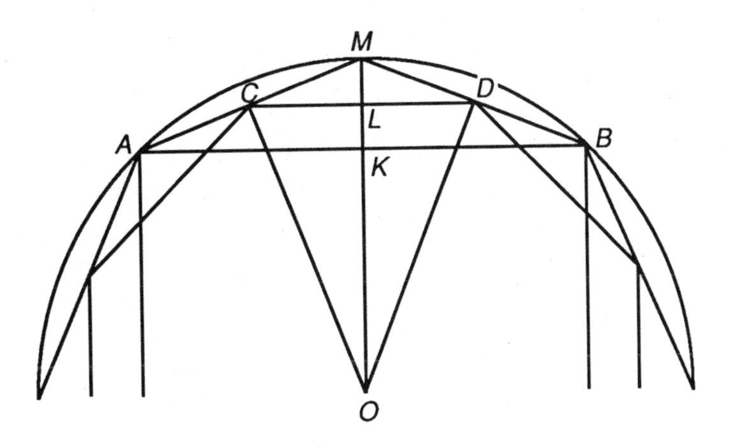

Consider a regular m-gon P_1 with side $\overline{AB}$ centered at 0. Let K and M denote the respective midpoints of segment $\overline{AB}$ and arc $\overset{\frown}{AB}$ of the circumscribed circle to P_1. Then:

$$\begin{aligned} m \cdot AB &= p &&= \text{perimeter of } P_1 \\ OK &= r_1 &&= \text{inradius of } P_1 \\ OM &= R_1 &&= \text{circumradius of } P_1. \end{aligned}$$

Now let C and D be the midpoints of $\overline{AM}$ and $\overline{MB}$, and consider the regular polygon P_2 of side length CD. Since $\Delta CDM \approx \Delta ABM$ and $CM = \frac{1}{2}AM$, we see that $CD = \frac{1}{2}AB$. Thus P_2 has twice the number of sides as P_1, but each is half as long, and so

$$\begin{aligned} 2m \cdot CD &= p &&= \text{perimeter of } P_2 \\ OL &= r_2 &&= \text{inradius of } P_2 \\ OC &= R_2 &&= \text{circumradius of } P_2. \end{aligned}$$

Next we show how to obtain r_2 and R_2, as they depend on r_1 and R_1. Since L is the midpoint of $\overline{KM}$ we see that $OL = \frac{1}{2}(OK + OM)$; that is, r_2 is the arithmetic mean of r_1 and R_2, namely

$$r_2 = \frac{1}{2}(r_1 + R_1). \tag{1}$$

The similar right triangles OLC and OCM tell us that $OL/OC = OC/OM$; that is, $r_2/R_2 = R_2/R_1$. Solving for R_2 we have

$$R_2 = \sqrt{r_2 R_1}, \tag{2}$$

which shows us that R_2 is the geometric mean of r_2 and R_1.

The transformation from the m-gon P_1 to the 2m-gon P_2 has preserved the perimeter p and changed the inradius and circumradius according to equations (1) and (2). Repeated applications of the transformation generate a sequence of polygons $P_1, P_2, \cdots, P_n, P_{n+1}, \cdots$, each of perimeter p. The inradii and circumradii transform according to the equations

$$r_{n+1} = \frac{1}{2}(r_n + R_n) \tag{3}$$

$$R_{n+1} = \sqrt{r_{n+1} R_n}. \tag{4}$$

Let us now see how (1) and (2) can be used recursively to find approximations to π. The incircle to P_n has circumference $2\pi r_n$, while the circumcircle has circumference $2\pi R_n$. Also P_n itself has perimeter p, and so $2\pi r_n < p < 2\pi R_n$. This inequality can be rearranged to trap π; we see that

$$\frac{p}{2R_n} < \pi < \frac{p}{2r_n}. \tag{5}$$

As n increases, the estimate of π given by (5) improves.

The sequence $\left\{\frac{p}{2R_n}\right\}$ increases to the limit π, while the sequence $\{\frac{p}{2r_n}\}$ decreases to the same limit π.

Example.

Suppose we begin with a square P_1 of perimeter 8. Then $r_1 = 1$ and $R_1 = \sqrt{2}$, so we get $\dfrac{8}{2\sqrt{2}} < \pi < \dfrac{8}{2}$, or $2\sqrt{2} < \pi < 4$.

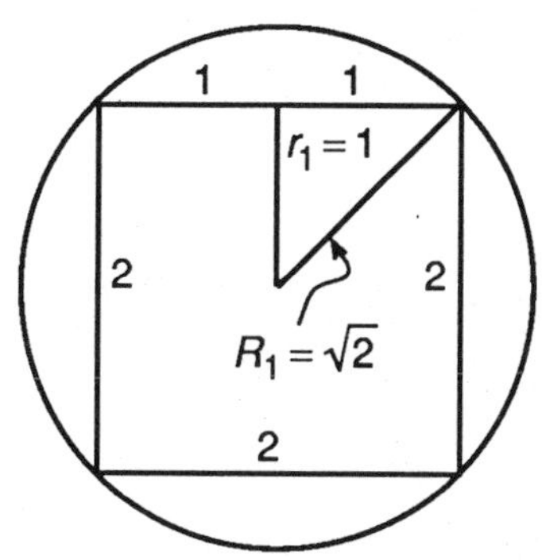

To obtain an improved value we use equations (1) and (2) in succession, finding

$$r_2 = \frac{1}{2}(1+\sqrt{2}) = 1.207\cdots,$$

$$R_2 = \sqrt{\frac{1}{2}(1+\sqrt{2})\sqrt{2}} = 1.306\cdots,$$

then $\dfrac{8}{2R_2} = \dfrac{8}{2(1.306\cdots)} = 3.061\cdots, \dfrac{8}{2r_2} = \dfrac{8}{2(1.207\cdots)} = 3.313\cdots,$

and so $3.061\cdots < \pi < 3.313\cdots$

Exercises on Evaluating π

1. For P_1 the square of perimeter 8, calculate r_3 and R_3 and obtain an estimate of π.

2. Suppose P_1 is a hexagon of circumference 6. Show that $r_1 = \sqrt{3}/2$ and $R_1 = 1$, to obtain $3 < \pi < 2\sqrt{3}$. Now use equations (3), (4), and (5), with $n = 1, 2, 3$ to improve your estimate of π. If a programmable calculator or computer is available, see what estimate is obtained at $n = 10$.

3. Show by induction that $R_{n+1}^2 - r_{n+1}^2 = \frac{1}{4^n}(R_1^2 - r_1^2)$.

Answers to Exercises on Evaluating π

1. $r_3 = \frac{1}{2}(1.207\ldots + 1.306\ldots) = 1.256\ldots$

$$R_3 = \sqrt{(1.256\ldots)(1.306\ldots)} = 1.281\ldots$$

$$\frac{8}{2R_3} = 3.121\ldots < \pi < 3.182\ldots = \frac{8}{2r_3}.$$

2. Since ΔOAB is equilateral, it is clear that $r_1 = \sqrt{3}/2$ and $R_1 = 1$, and so $\frac{6}{2R_1} = 3 < \pi < 2\sqrt{3} = \frac{6}{2r_1}$.

	r_n	R_n	$\frac{p}{2R_n}$	$\frac{p}{2r_n}$
n=1	0.866	1.000	3.000	3.464
2	0.933	0.966	3.106	3.215
3	0.949	0.958	3.133	3.160

At $n = 10$, one finds $\pi \approx 3.141592\ldots$, accurate to six decimal places.

3. For $n = 1$ we have $R_2^2 - r_2^2 = r_2R_1 - r_2^2$

$$= r_2(R_1 - r_2) = \left[\frac{1}{2}(r_1 + R_1)\right]\left(R_1 - \frac{1}{2}(r_1 + R_1)\right)$$

$$= \frac{1}{2}(r_1 + R_1)\left(-\frac{1}{2}r_1 + \frac{1}{2}R_1\right) = \frac{1}{4}(R_1^2 - r_1^2).$$

Let $k \geq 1$, and suppose $R_{k+1}^2 - r_{k+1}^2 = \frac{1}{4^k}(R_1^2 - r_1^2)$.

Then $R_{k+2}^2 - r_{k+2}^2 = r_{k+2}R_{k+1} - r_{k+2}^2$

$$= r_{k+2}\left(R_{k+1} - r_{k+2}\right) = \left[\frac{1}{2}(r_{k+1} + R_{k+1})\right]\left[R_{k+1} - \frac{1}{2}(r_{k+1} + R_{k+1})\right]$$

$$= \frac{1}{4}(R_{k+1}^2 - r_{k+1}^2) = \frac{1}{4}\cdot\frac{1}{4^k}(R_1^2 - r_1^2) = \frac{1}{4^{k+1}}(R_1^2 - r_1^2),$$

which completes the induction.

Euler's Constant

A speaker before a student audience said, "Let God choose three numbers at random," and wrote

$$
\begin{array}{l}
3.14159 \;\; 26535 \cdots \\
2.71828 \;\; 18284 \cdots \\
0.57721 \;\; 56649 \cdots
\end{array}
$$

A member of the audience asked, "What is that third number? I recognize the first two, but not the third." The student had recognized π and e, but was not yet acquainted with Euler's constant $\gamma = 0.57721\cdots$ Since γ arises naturally in numerous diverse contexts, including number theory and probability, there is considerable value in studying "The Third Constant."

Euler discovered γ in the mid 1730s, when he wished to compare $h_n = 1 + \frac{1}{2} + \cdots + \frac{1}{n}$ (the nth partial sum of the harmonic series) with $\ln(n+1)$. Both can be viewed as areas: h_n is the sum of areas of rectangles, and $\ln(n+1) = \int_1^{n+1} \frac{1}{x} dx$ is the area under the $y = \frac{1}{x}$ curve, as shown below.

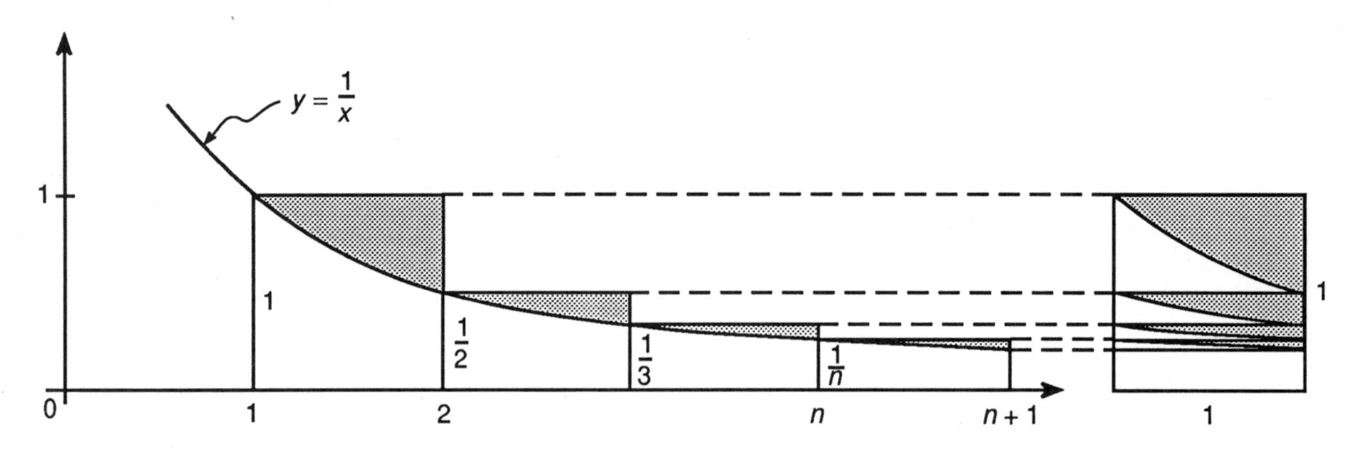

Let $a_n = 1 + \frac{1}{2} + \frac{1}{3} + \cdots + \frac{1}{n} - \ln(n+1)$ denote the difference between these areas, represented by the sum of the shaded triangle-like areas in the upper right corners of the rectangles. By placing these regions in a unit square we see that

$$a_1 < a_2 < \cdots < a_n < a_{n+1} < 1.$$

Thus $\{a_n\}$ is a bounded monotone increasing sequence, and so it has a limit. Euler defined the constant γ as the limit of the sequence $\{a_n\}$; that is,

$$\gamma = \lim_{n \to \infty} [(1 + \frac{1}{2} + \frac{1}{3} + \cdots + \frac{1}{n}) - \ln(n+1)].$$

Since $a_{30} = 1 + \frac{1}{2} + \cdots + \frac{1}{30} - \ln 31 = 0.56099\cdots$, we know that $0.56099 < \gamma$.

To estimate γ from above, let us first notice that $a_n = h_n - \ln(n+1) = h_n - \ln n - \ln(\frac{n+1}{n})$. Since $\lim_{n \to \infty} \ln(\frac{n+1}{n}) = \ln 1 = 0$, we see that we also have

$$\gamma = \lim_{n \to \infty} [(1 + \frac{1}{2} + \cdots + \frac{1}{n}) - \ln n].$$

By the "squeezing principle," we conclude that the sequence $\{b_n\}$, where

$$b_n = 1 + \frac{1}{2} + \cdots + \frac{1}{n} - \ln(n + \frac{1}{2}),$$

also converges to γ. Since $b_1 = 0.5945\cdots, b_2 = 0.5837\cdots, b_3 = 0.5805, \cdots$ one might hope that the sequence $\{b_n\}$ monotonely decreases to γ. Let us check this out.

Since

$$\begin{aligned} b_n - b_{n+1} &= [h_n - \ln(n + \tfrac{1}{2})] - [h_{n+1} - \ln(n + 1 + \tfrac{1}{2})] \\ &= -\tfrac{1}{n+1} - \ln(n + \tfrac{1}{2}) + \ln(n + \tfrac{3}{2}), \end{aligned}$$

let us consider the function

$$
\begin{aligned}
f(x) &= -\tfrac{1}{x+1} - \ln(x+\tfrac{1}{2}) + \ln(x+\tfrac{3}{2}) \\
&= -\tfrac{1}{x+1} + \ln\tfrac{x+\frac{3}{2}}{x+\frac{1}{2}},
\end{aligned}
$$

defined on $x \geq 1$. We see that $\lim_{x\to\infty} f(x) = -0 + \ln 1 = 0$, and calculate that

$$
\begin{aligned}
f'(x) &= \frac{1}{(x+1)^2} - \frac{1}{x+\frac{1}{2}} + \frac{1}{x+\frac{3}{2}} \\
&= \frac{1}{(x+1)^2} - \frac{1}{x^2+2x+\frac{3}{4}} \\
&= \frac{1}{x^2+2x+1} - \frac{1}{x^2+2x+\frac{3}{4}} < 0.
\end{aligned}
$$

Thus f decreases asymptotically to 0 on $x \geq 1$, meaning $f(x) > 0$ for all $x \geq 1$. In particular, $f(n) = b_n - b_{n+1} > 0$, and so

$$b_1 > b_2 > \cdots > b_n > b_{n+1} > \cdots > \gamma.$$

Since $b_{30} = 1 + \frac{1}{2} + \cdots + \frac{1}{30} - \ln(30 + \frac{1}{2}) = 0.57726\cdots$ we see that $\gamma < 0.57726\cdots$

Combining our upper and lower estimates, we see that

$$0.56099\cdots < \gamma < 0.57726\cdots$$

Although γ can be estimated with unlimited precision, little else is known about this "third constant." It is not even known whether or not γ is rational.

Computations with Euler's Constant

1. We have seen that $a_n = (1 + \frac{1}{2} + \frac{1}{3} + \cdots + \frac{1}{n}) - \ln(n+1)$ is represented by a shaded area in the unit square. γ is approached by filling in some, but not all, of the $1 \times \frac{1}{n+1}$ rectangle on the bottom of the unit square.

   (a) Explain why $a_n < \gamma < a_{n+1} + \frac{1}{n+1}$.

   (b) Take $n = 10$ and $n = 20$, and with the use of a calculator verify the respective estimates

$$0.531\cdots < \gamma < 0.621\cdots$$

$$0.553\cdots < \gamma < 0.600\cdots$$

2. Fill in the numerical values in the table below, using calculators (programmable ones if available) or computers. Recall that $h_n = 1 + \frac{1}{2} + \cdots + \frac{1}{n}$, $a_n = h_n - \ln(n+1)$, $b_n = h_n - \ln(n + \frac{1}{2})$, and $\gamma = 0.5772156649\cdots$.

	h_n	a_n	$\gamma - a_n$	b_n	$b_n - \gamma$
n = 1	1.00000	0.306853	0.270363	0.594535	0.017319
2	1.50000	0.401388	0.175828	0.583709	0.006494
3	1.83333	0.447039	0.130177	0.580570	0.003355
4	2.08333	0.473895	0.103320	0.579256	0.002040
5	2.28333	0.491574	0.085642	0.578585	0.001370
10	2.92897	0.531073	0.046143	0.577593	0.000377
20	3.597740	0.553217	0.023998	0.577315	0.000099

3. Since $a_n = h_n - \ln(n+1) < \gamma < h_n - \ln(n + \frac{1}{2}) = b_n$ we see $\gamma + \ln(n + \frac{1}{2}) < h_n < \gamma + \ln(n+1)$. Use a calculator to estimate h_{100} and h_{500}.

Reference: D. Bushaw and S. Saunders, "The Third Constant," *Northwest Science 59* (1985), 147-158.

The Big Picture

Series

Would you agree that $1+\frac{1}{2}+\frac{1}{4}+\frac{1}{8}+\frac{1}{16}+\cdots$ should be 2 if it is anything? In fourth grade you wrote $1/3 = .333\cdots$. Do you still think that $1/3 = 3/10+3/100+3/1000+\cdots$? Why? Assume $a_1, a_2, \cdots, a_n, \cdots$ is a sequence of numbers. We want to make sense, when we can, of the sum of the sequence $a_1 + a_2 + \cdots$, which we shorten by writing $\sum_{n=1}^{\infty} a_n$.

Why should $1+\frac{1}{2}+\frac{1}{4}+\cdots = \sum_{n=1}^{\infty} \frac{1}{2^{n-1}} = 2$? Note what happens as we continue to add.

one term	$a_1 = 1$. Call it s_1 for first "partial sum."
two terms	$a_1 + a_2 = 1 + \frac{1}{2} = \frac{3}{2}$. Call it s_2 for second partial sum.
three terms	$a_1 + a_2 + a_3 = 1 + \frac{1}{2} + \frac{1}{4} = \frac{7}{4}$. Call it s_3.
$\vdots$	(there is a pattern)
	$a_1 + a_2 + \cdots + a_n = 1 + \frac{1}{2} + \cdots + \frac{1}{2^{n-1}} = \frac{2^n - 1}{2^{n-1}} = 2 - \frac{1}{2^{n-1}}$. Call this s_n.

Note that $\lim_{n\to\infty} s_n = 2$ where the limit is our sequence limit that we have already developed.

We are ready for a definition.

> Set $s_n = a_1 + a_2 + \cdots + a_n$. Then $\sum_{n=1}^{\infty} a_n = L$ if and only if the sequence $s_1, s_2, \cdots$ converges to L.

Finally we know what we did in the fourth grade! Form the partial sum sequence

$$\begin{aligned} s_1 &= 3/10 \\ s_2 &= 33/100 \\ s_3 &= 333/1000 \end{aligned}$$

and the sequence $s_1, s_2, s_3, \cdots$ converges to $1/3$. Thus $1/3 = .333\cdots$.

Some important facts are:

1. If $\sum_{1}^{\infty} a_n$ converges, then the sequence $\{a_n\}$ has to converge to zero.

2. The sequence $\{a_n\}$ can converge to zero, yet $\sum_{n=1}^{\infty} a_n$ not converge. (Remember the harmonic series $\sum_{n=1}^{\infty} \frac{1}{n}$ diverges even though $\{\frac{1}{n}\} \to 0$.)

3. If the a_n's are all positive (or negative), then the sequence of partial sums is monotone, and hence converges if and only if it is bounded.

4. From (3) we get common sense Comparison Tests. One example is: If $0 \leq b_n \leq a_n$ and $\sum_{n=1}^{\infty} a_n$ converge, then $\sum_{n=1}^{\infty} b_n$ converges. (The only way for $\sum_{n=1}^{\infty} b_n$ to diverge is for its partial sum sequence to grow without bound. That is impossible. The series $\sum_{n=1}^{\infty} a_n$ has a partial sum sequence whose terms are at least as big, and they don't grow without bound because we are given $\sum_{n=1}^{\infty} a_n$ convergent.) Similarly, if $\sum_{n=1}^{\infty} b_n$ diverges, then so does $\sum_{n=1}^{\infty} a_n$.

In general it is hard to find the actual sum (or limit) of a convergent series. We can determine that many series converge without knowing what the limit is. A number of tests will show us how to do that. One notable and important exception to that statement, as we will see, is a geometric series whose sum is easy to find.

Summary of Results for Series Having Both Positive and Negative Terms

The tests that have been developed for series with positive terms (ratio test, root test, comparison test, integral test, and so on) can be applied twice to test some mixed term series for convergence or divergence. Suppose the series $a_1 + a_2 + a_3 + \cdots$ has both positive and negative terms. The series we get by taking only the positive terms is called the *positive part* of the series; the series of negative terms is the *negative part*.

Example: $1 - 1 + \frac{1}{2} - \frac{1}{2} + \frac{1}{4} - \frac{1}{3} + \frac{1}{8} - \frac{1}{4} + \frac{1}{16} - \frac{1}{5} + \cdots$ has positive part $1 + \frac{1}{2} + \frac{1}{4} + \frac{1}{8} + \cdots$ and negative part $-1 - \frac{1}{2} - \frac{1}{3} - \frac{1}{4} - \frac{1}{5} - \cdots$. Rearrangements of the original series such as $1 + \frac{1}{2} - 1 - \frac{1}{2} + \frac{1}{4} + \frac{1}{8} - \frac{1}{3} - \frac{1}{4} + \cdots$ would have the same positive and negative parts as the first series.

The following facts summarize the convergence properties of series having both positive and negative terms.

1. If $\lim_{n\to\infty} a_n$ is not zero, the series and any of its rearrangements will diverge. This applies to all series whether or not there are mixed signs.

In case 1, look no further. (Caution: A common error is to conclude that a series converges if $\lim_{n\to\infty} a_n = 0$. That is a necessary but not sufficient condition for convergence. Remember the harmonic series, and don't make this error!)

2. If both positive and negative parts converge, then $\sum a_n$ converges; in fact, $\sum |a_n|$ converges also.

That is, the series is absolutely convergent. The reason is easy to see; the partial sums for $\sum |a_n|$ are bounded and monotone in this case, and thus converge. But we know any absolutely convergent series converges itself, so $\sum a_n$ converges. Another important property of these series is that *any* rearrangement of them will still converge to the same limit. This is a super-extension of the commutative property $a + b = b + a$.

Example: $1 - 1 + \frac{1}{2} - \frac{1}{4} + \frac{1}{4} - \frac{1}{9} + \cdots$ converges since both $1 + \frac{1}{2} + \frac{1}{4} + \cdots$ and $-1 - \frac{1}{4} - \frac{1}{9} - \cdots$ are convergent series. The series $1 + \frac{1}{2} - 1 + \frac{1}{4} + \frac{1}{8} - \frac{1}{4} + \frac{1}{16} + \frac{1}{32} - \frac{1}{9} + \cdots$ would converge to the same limit, since it is a rearrangement of the original, absolutely convergent series.

3. If the positive part converges and the negative part diverges, the series diverges because the partial sums approach $-\infty$.

This is also true of any rearrangement of the series.

Example: $1-1+\frac{1}{2}-\frac{1}{2}+\frac{1}{4}-\frac{1}{3}+\frac{1}{8}-\frac{1}{4}+\frac{1}{16}-\frac{1}{5}+\cdots$ diverges, since $1+\frac{1}{2}+\frac{1}{4}+\frac{1}{8}+\cdots$ converges but $-1-\frac{1}{2}-\frac{1}{3}-\frac{1}{4}-\cdots$ diverges. Note that $\lim_{n\to\infty} a_n = 0$. It is easy to see why partial sums for the series approach $-\infty$. As more terms are added, the positive contribution approaches 2, while the negative contribution approaches $-\infty$. The net result, of course, is that the partial sums approach $-\infty$.

> 4. If the positive part diverges and the negative part converges, the series diverges because the partial sums approach $+\infty$.

This is also true of any rearrangement of the series. This is just case 3 above with all the signs of the series changed.

> 5. If both positive and negative parts diverge and $\lim_{n\to\infty} a_n = 0$, nothing definite can be said until further tests are run. If it converges we say it converges conditionally.

Thus a conditionally convergent series is one that converges but does not converge absolutely.

Examples: $1-\frac{1}{2}+\frac{1}{3}-\frac{1}{4}+\frac{1}{5}+\cdots$ converges conditionally to $\ln 2$. Both positive and negative parts diverge, so the absolute series diverges. Yet rearrangements of this series can have very different behaviors.

a. *A rearrangement can converge to $\pi/2$ (or any other finite number).* It is easy to describe how to get such a rearrangement. Use positive terms until you first exceed $\pi/2$. In this case start with $1+\frac{1}{3}+\frac{1}{5}+\frac{1}{7}$, which is greater than $\pi/2$. Then add negative terms until sums are first less than $\pi/2$, in this case $1+\frac{1}{3}+\frac{1}{5}+\frac{1}{7}-\frac{1}{2} < \pi/2$. Then add more new positive terms until we first exceed $\pi/2$, then negative terms until the sum is first less than $\pi/2$, and so on.

To see that this works we ask two questions: 1. Can I always add enough terms to get back across $\pi/2$ regardless which way I am going? (Yes, the positive part approached $+\infty$, the negative part $-\infty$, and at any stage only a finite amount of either has been used. Plenty remains to carry us back across $\pi/2$.) 2. Will the series that oscillates back and forth across $\pi/2$ actually converge to $\pi/2$, or could the jumps across continue to be large enough so as not to converge to $\pi/2$? (Since $\lim_{n\to\infty} a_n = 0$ the amount we jump across $\pi/2$ goes to 0, so the series does converge to $\pi/2$.)

b. *A rearrangement can diverge to $+\infty$(or $-\infty$).*
Use the same idea as above: first add positive terms until you get to 1, throw in a negative term, then add enough positive terms to get to 2, throw in another

negative term, then add enough positive terms to get to 3, throw in another negative term, and so on.

c. *A rearrangement can diverge because partial sums oscillate but are nevertheless bounded.*
Use the same idea. Add positive terms until you first exceed 100, then enough negative terms until your sum is first less than -100, then repeat, and so on.

The preceeding examples show why nothing can be said about the convergence of a series whose positive and negative parts diverge and whose terms approach zero. There is one theorem that covers a very special case.

6. If $a_1 + a_2 + a_3 + a_4 + \cdots$ is an alternating series (that is $a_1, a_3, a_5 \cdots$ are all positive while $a_2, a_4, a_6, \cdots$ are all negative, so signs alternate) and $\|a_1\| \geq \|a_2\| \geq \|a_3\| \cdots$ and $\lim_{n\to\infty} a_n = 0$, then the series converges.

Example: $\frac{1}{100} - \frac{1}{103} + \frac{1}{106} - \frac{1}{109} + \cdots$ converges.

Series

For each of the questions fill the blank or blanks with a response from the following list:

(a) diverges to $-\infty$

(b) diverges to $+\infty$

(c) diverges but partial sums are bounded

(d) converges conditionally

(e) converges absolutely

(f) might converge, might diverge

1. If the negative part of a series diverges while the positive part converges, then the series ____________.

2. If the negative terms of the series in Exercise 1 are halved and the positive terms are all doubled, then the new series ____________.

3. If both the negative and positive parts of a series converge, then the series ____________.

4. If the positive terms of the series in Exercise 3 are all doubled and the negative terms all halved, the new series ____________.

5. If $a_{2n} > 0$, $a_{2n-1} < 0$, $\mid a_{2n-1} \mid = 2 \mid a_{2n} \mid$ (that is, the negative terms are twice the size in absolute value as the following positive term) for all n, then the series $\sum a_n$ ____________ or ____________.

6. If both the positive and negative parts of the series diverge and $a_n \to 0$ as $n \to \infty$, then the series $\sum a_n$ ____________.

7. If both the positive and negative parts of the series diverge, if $a_n \to 0$, $a_{2n} < 0, a_{2n-1} > 0$ and $\mid a_n \mid > \mid a_{n+1} \mid$ for all n, then the series ____________.

8. If the terms of the series in Exercise 1 are rearranged, the new series ____________.

9. If the terms of the series in Exercise 3 are rearranged, the new series ____________.

10. If the terms of the series in Exercise 7 are rearranged, the new series ____________.

11. If $\mid a_n \mid > \mid a_{n+1} \mid$, $a_{2n} < 0$, $a_{2n-1} > 0$ for all n, yet $\lim a_n \neq 0$, then the series ____________.

Answers to Series Probes

1. a.

2. a. The negative part still diverges, and the positive part still converges.

3. e.

4. e. Positive and negative parts still both converge.

5. e, if the positive (therefore negative) part converges, or a. if the positive (therefore negative) part diverges. In the latter case partial sums diverge to $-\infty$ because the positive part grows only half as fast as the negative part.

6. f. $\sum(-1)^n \frac{1}{n}$ converges, while $1 - 2 + \frac{1}{2} - 1 + \frac{1}{3} - \frac{2}{3} + \frac{1}{4} - \frac{2}{4} + \cdots$ diverges.

7. d. This is one of the standard theorems.

8. a. The negative part still diverges, while the positive part still converges.

9. e.

10. f. See example under rule 5 on page 158.

11. c.

Alternating Series: The Confused Frog

A frog hops successive positive distances $a_1, a_2, \cdots, a_n, \cdots$ but he is: (1) increasingly tired, so that $a_1 > a_2 > \cdots > a_n$ and $a_n \to 0$; and (2) confused (his first hop is to the right, second to the left, third to the right, and so on.)

Let us draw the hopping frog's positions:

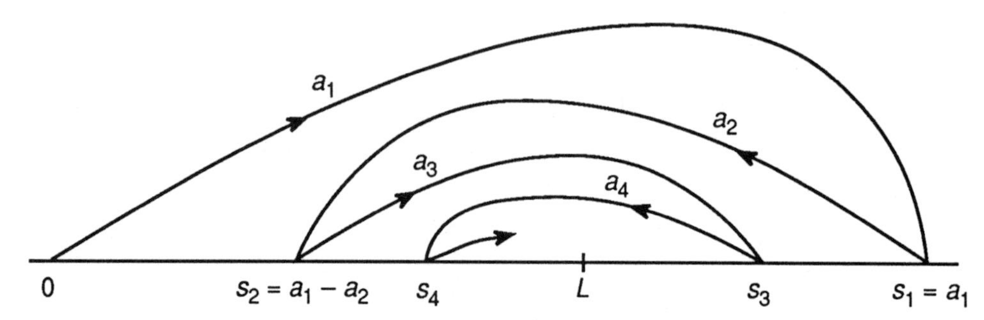

Let s_n = x-coordinate of frog's position after n hops.

$$\begin{aligned} s_1 &= a_1 \\ s_2 &= a_1 - a_2 \\ &\vdots \\ s_n &= a_1 - a_2 + \cdots - (-1)^n a_n. \end{aligned}$$

Exercises on the Confused Frog

1. What can be said about the sequences of positions

$$s_1, s_3, s_5, \cdots, s_{2n+1}, \cdots?$$

$$s_2, s_4, s_6, \cdots, s_{2n}, \cdots?$$

2. Why do you know $s_1, s_3, \cdots$ and $s_2, s_4, \cdots$ have a common limit L?

3. Notice the frog always hops across L on every hop. If the frog is at $s_{19} = 75$ and $a_{20} = 5$ and $a_{21} = 3$, what interval is guaranteed to contain L?

Answers to Exercises on the Confused Frog

1. The sequence $s_1, s_3, \cdots, s_{2n+1}, \cdots$ decreases and $s_2, s_4, \cdots, s_{2n}, \cdots$ increases.

2. $0 < s_2 < s_4 \cdots < s_3 < s_1$
   The sequences (s_{2n}) and (s_{2n+1}) are bounded and monotone, so have limits; say, $L = \lim s_{2n}$ and $M = \lim s_{2n-1}$. Since $s_{2n+1} - s_{2n} = a_{2n+1}$ and $\lim a_n \to 0$, we see (taking limits) $M - L = 0$.

3. $L \in (70, 73)$.

Is e a Rational Number?

The Pythagoreans of ancient Greece based their philosophy on the concept of number–more precisely, whole numbers. One of the earliest crises to occur in mathematical history was the discovery (by the Pythagoreans themselves!) that not all numbers were expressible as a ratio of whole numbers. For example, it is not possible to find integers p and q for which p/q is equal to $\sqrt{2}$, and so $\sqrt{2}$ is irrational.

It is not always easy to know if a given number is rational or not. For example, it was not known until the middle of the eighteenth century that π is irrational, as proved by J. H. Lambert (1728-1777). Euler showed the e is irrational, but he didn't settle the case of "the third constant" $\gamma = \lim_{n\to\infty}[(1+\frac{1}{2}+\cdots+\frac{1}{n}) - \ln\, n] = 0.577\ldots$. Indeed, no one yet knows whether or not γ is rational.

In the sequence of exercises that follow we show that the assumption that e is a rational number, say $e = \frac{p}{q}$, leads to a contradiction. We assume it is known that

$$e^x = \sum_{n=0}^{\infty} \frac{x^n}{n!} \qquad \text{all } x,$$

so that with the choice $x = -1$ we have

$$\frac{1}{e} = \sum_{n=0}^{\infty} \frac{(-1)^n}{n!}. \tag{1}$$

Exercises on: Is e a Rational Number?

1. Recall that the sum of an alternating series is between any two successive partial sums. Use equation (1) to show that $\frac{1}{e} \in \left(\frac{1}{3}, \frac{1}{2}\right)$.

2. If $\frac{1}{e} = \frac{q}{p}$, explain why $p > 2$.

3. If $\frac{1}{e} = \frac{q}{p}$, use (1) to show that

(a)
$$\frac{q}{p} - \sum_{n=0}^{p} \frac{(-1)^n}{n!} = \sum_{n=p+1}^{\infty} \frac{(-1)^n}{n!}$$

4. Multiply the result of question 3 by $(-1)^{p+1}p!$ to get

(b)
$$(-1)^{p+1}\left[q(p-1)! - \sum_{n=0}^{p}(-1)^n \frac{p!}{n!}\right] = \sum_{n=p+1}^{\infty} \frac{p!}{n!}(-1)^{n+p+1}$$

5. Explain why the left-hand side of (b) is an integer.

6. Explain why the sum of the series on the right hand side of (b) is in the interval $\left(\frac{1}{p+2}, \frac{1}{p+1}\right)$.

7. Your assumption that e is a rational number p/q has led to a contradiction. Explain.

Answers to Exercises on: Is e a Rational Number?

1. The first four partial sums of $\frac{1}{e} = \sum_{n=0}^{\infty} \frac{(-1)^n}{n!} = 1 - 1 + \frac{1}{2!} - \frac{1}{3!} + \cdots$ are $1, 0, \frac{1}{2}, \frac{1}{3}$. Successive partial sums always cross over the sum of an alternating series, and thus $\frac{1}{e} \in \left(\frac{1}{3}, \frac{1}{2}\right)$.

2. No fractions of the form $\frac{q}{2}$ belong to $\left(\frac{1}{3}, \frac{1}{2}\right)$, so $p > 2$.

3. Straightforward rearrangement of series.

4. Straightforward rearrangement of series.

5. For $0 \le n \le p$, we have $\frac{p!}{n!} = p(p-1)\cdots(p-n+1)$, and so each term of the left side of (b) is an integer.

6. The right side of (b) is the convergent alternating series $\frac{1}{p+1} - \frac{1}{(p+2)(p+1)} + \frac{1}{(p+3)(p+2)(p+1)} - \cdots$.
   The first two partial sums are $\frac{1}{p+1}$ and $\frac{1}{p+1} - \frac{1}{(p+2)(p+1)} = \frac{1}{p+2}$, so the sum of the series is between $\frac{1}{p+2}$ and $\frac{1}{p+1}$.

7. Since $\left(\frac{1}{p+2}, \frac{1}{p+1}\right)$, for $p > 2$, is an interval that contains no integers, equation (b) is never valid when p and q are integers.

Infinite Series and Population Growth

Many of the problems in today's world, such as hunger, pollution, and poverty are related to the rapid growth in human population. The statistical study of populations, called *demography*, makes heavy use of mathematical methods, often calculus methods, to provide tools for analysis. By making some simplifying assumptions, we can show how this is done.

A key indicator of the population growth rate is *total fertility rate*, which is defined as the average number of children a woman will have. If the rate is two, the woman and her husband have exactly replaced themselves and the population is steady. If the rate exceeds two there will be population growth.*

We will now consider two examples, with the (somewhat inaccurate) assumption that boy and girl children are equally likely occurrences in any birth.

Example 1. Let us assume every couple in a country continues to have children until a boy is born, and then has no more children. This situation is nicely visualized by a tree diagram.

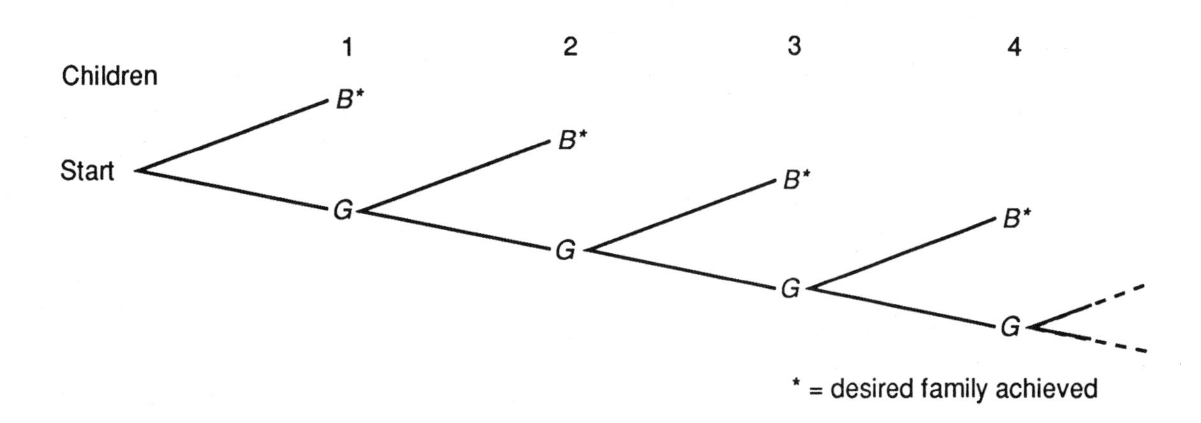

We see that $\frac{1}{2}$ of the women have one child, $\frac{1}{4}$ of the women have a girl and then a son, $\frac{1}{8}$ of the women have two girls and then a son, and so on. The average number of children is then given by

$$\frac{1}{2}(1) + \frac{1}{4}(2) + \frac{1}{8}(3) + \cdots + \frac{1}{2^n}(n) + \cdots.$$

*Taking child mortality into account, the replacement rate is actually about 2.1 in developed countries and about 2.5 in underdeveloped countries.

To sum this series, we consider the partial sums

$$\begin{aligned}
s_1 &= \tfrac{1}{2} && = 2 - \tfrac{3}{2} \\
s_2 &= \tfrac{1}{2} + \tfrac{2}{4} = \tfrac{4}{4} && = 2 - \tfrac{4}{4} \\
s_3 &= \tfrac{1}{2} + \tfrac{2}{4} + \tfrac{3}{8} = \tfrac{11}{8} && = 2 - \tfrac{5}{8} \\
s_4 &= \tfrac{1}{2} + \tfrac{2}{4} + \tfrac{3}{8} + \tfrac{4}{16} = \tfrac{26}{16} && = 2 - \tfrac{6}{16}.
\end{aligned}$$

As induction can verify, we have $s_n = 2 - \dfrac{n+2}{2^n}$. Letting $n \to \infty$, and using the fact that $(n+2)/2^n \to 0$, we find that the fertility rate is $\lim s_n = 2$. Of course, since families with more than 12 children are exceedingly rare, the actual rate is smaller than 2.

Remark. Another way to sum the series makes use of the infinite geometric series

$$\frac{1}{1-x} = 1 + x + x^2 + \cdots, \quad |x| < 1.$$

Since this power series can be differentiated term-by-term, we also have

$$\frac{1}{(1-x)^2} = 1 + 2x + 3x^2 + \cdots.$$

Multiplying by x and then setting $x = \frac{1}{2}$ gives

$$\frac{(1/2)}{(1-1/2)^2} = 1(\frac{1}{2}) + 2(\frac{1}{2})^2 + 3(\frac{1}{2})^3 + \cdots,$$

which is equivalent to the desired result:

$$2 = \frac{1}{2}(1) + \frac{1}{4}(2) + \frac{1}{8}(3) + \cdots.$$

Example 2. Let us assume each couple in a certain culture desires both a son and a daughter. Each woman has children until she has a child of the opposite sex from the first child, and then has no more.

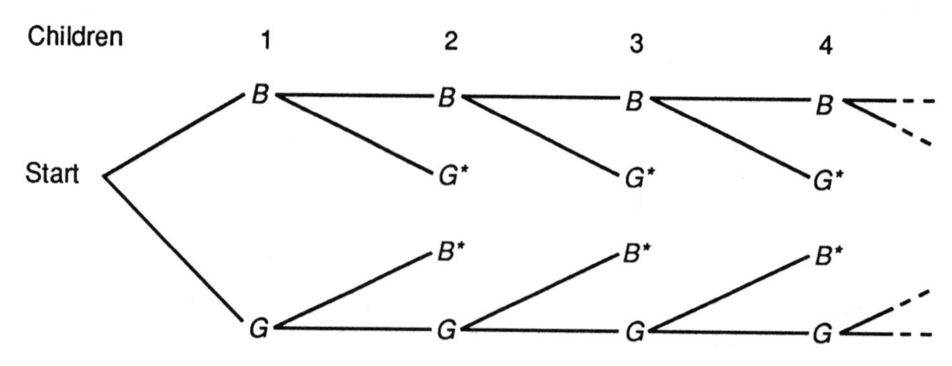

As the tree diagram shows, no women have a single child, one-half the women have two children, one-quarter have three children, one-eighth have four children, and so forth. The average number of children is given by the infinite series

$$\frac{1}{2}(2) + \frac{1}{4}(3) + \frac{1}{8}(4) + \cdots + \frac{1}{2^n}(n+1) + \cdots.$$

The partial sums are

$$s_1 = 1 = 3 - \tfrac{4}{2}$$

$$s_2 = \tfrac{7}{4} = 3 - \tfrac{5}{4}$$

$$s_3 = \tfrac{18}{8} = 3 - \tfrac{6}{8}$$

and, as may be checked by mathematical induction,

$$s_n = 3 - \frac{1}{2^n}(n+3).$$

Thus the fertility rate is $\lim_{n\to\infty} s_n = 3$, which is above replacement level.

Remark. The series for example 2 can also be evaluated by using the geometric series formula $1 + 2x + 3x^2 + \cdots = 1/(1-x)^2$. Setting $x = \frac{1}{2}$ and subtracting 1 from both sides, we have

$$2 \cdot (\frac{1}{2}) + 3(\frac{1}{2})^2 + \cdots + (n+1)(\frac{1}{2})^n + \cdots = \frac{1}{(1-\frac{1}{2})^2} - 1 = 3.$$

Exercises on Infinite Series and Population Growth

Consider a culture in which each woman has children until two girls are born.

1. Draw a tree diagram that illustrates the situation.

2. Show that the average number of children for each woman is given by the infinite series
$$\frac{1}{4}(2) + \frac{2}{8}(3) + \frac{3}{16}(4) + \cdots + \frac{n}{2^{n+1}}(n+1) + \cdots.$$

3. Show that the partial sums are $s_n = 4 - \dfrac{(n+2)(n+3)+2}{2^{n+1}}$.

4. Show that the fertility rate is 4.

5. Differentiation of the infinite geometric series $\frac{1}{1-x} = 1 + x + x^2 + \cdots$ twice shows that $2 \cdot 1 + 3 \cdot 2x + 4 \cdot 3x^2 + \cdots + (n+1)nx^{n-1} + \cdots = 2/(1-x)^3$. Multiply by x^2 and then set $x = \frac{1}{2}$ to obtain the fertility rate in a different way.

Reference: Richard H. Schwartz, "Population Growth, Tree Diagrams, and Infinite Series," *The UMAP Journal* 6, No. 1 (1985), 35-40.

Answers to Exercises on Infinite Series and Population Growth

1.

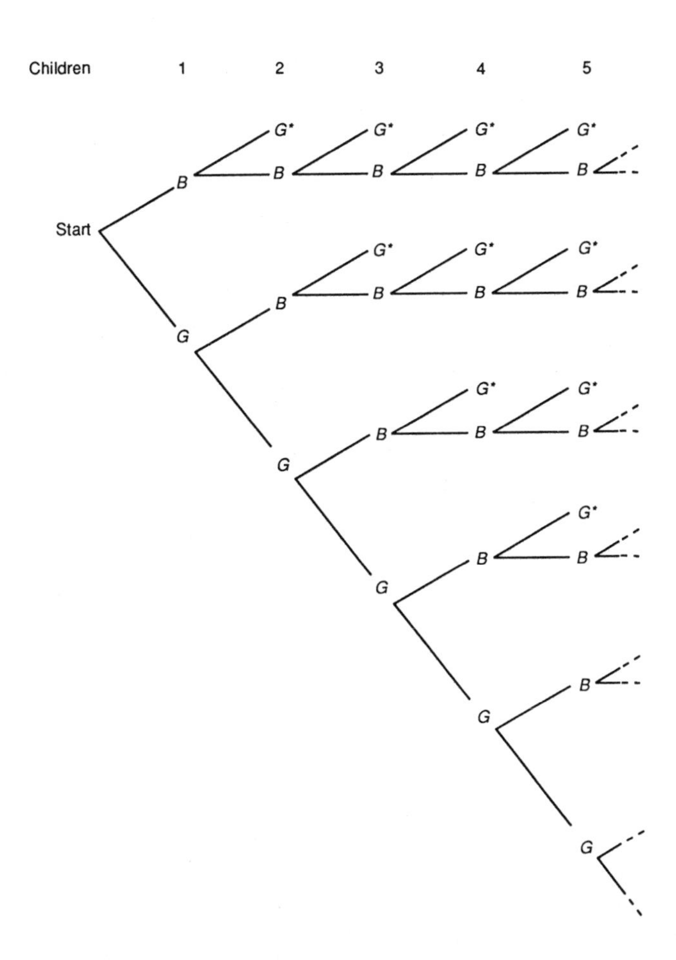

Note that the above column for $n+1$ children has n*'s.

2. Use $s_{n+1} = s_n + \frac{(n+1)}{2^{n+2}}(n+2)$ and mathematical induction.

3. $\lim s_n = 4$

4. $2 \cdot (\frac{1}{2})^2 + 3 \cdots (\frac{1}{2})^3 + \cdots = 2(\frac{1}{2})^2/(1-\frac{1}{2})^3 = 4.$

5. From $\frac{1}{1-x} = 1 + x + x^2 + \cdots$ we get $\frac{-1}{(1-x)^2} = 1 + 2x + 3x^2 + \cdots + (n+1)x^n + \cdots$ and $\frac{2}{(1-x)^3} = 2 \cdot 1 + 3 \cdot 2x + \cdots + (n+1)nx^{n-1} + \cdots$. Multiply by x^2 to get $\frac{2x^2}{(1-x)^3} = 2 \cdot 1x^2 + 3 \cdot 2x^3 + \cdots + (n+1)nx^{n+1} + \cdots$. Setting $x = \frac{1}{2}$ we get $4 = \frac{1}{4}(2) + \frac{2}{8}(3) + \cdots + \frac{n}{2^{n+1}}(n+1) + \cdots$.

The Divergence of the Harmonic Series: A Proof by Contradiction

Let us suppose the harmonic series is convergent, say with sum H, so that

$$H = 1 + \frac{1}{2} + \frac{1}{3} + \frac{1}{4} + \cdots.$$

By comparison, it would then follow that the series of odd terms and the series of even terms are also convergent. If D and E denote these sums, then we have

$$D = 1 + \frac{1}{3} + \frac{1}{5} + \cdots,$$

$$E = \frac{1}{2} + \frac{1}{4} + \frac{1}{6} + \cdots.$$

We easily see that

$$H = D + E \qquad \text{and} \qquad 2E = H,$$

and so $D = E$. This contradicts the obvious inequality $D > E$, which follows by term-by-term comparison of the series for D and E. The only alternative to our original supposition is that the harmonic series is divergent to infinity.

Remark. This proof (and other well-known proofs) of the divergence of the harmonic series are surprisingly simple in view of the slowness at which the partial sums creep off to infinity. If we sum over a subset S of the positive integers, it can be quite difficult to know if the series $\sum_{n \in S} \frac{1}{n}$ is convergent or divergent. For example, if $Z_i = \{n :$ n is a positive integer with exactly i zeros in its base ten numeral$\}$ then it can be shown that $\sum_{n \in Z_i} \frac{1}{n} \equiv \sigma_i$ is convergent. Indeed, it is known that $\sigma_0 \approx 23.10345$, and $\{\sigma_i\}_{i=0}^{\infty}$ is a decreasing sequence whose limit exceeds 19.28. Another example is to sum over prime numbers; it can be shown that in this case

$$\sum_{p = \text{ prime}} \frac{1}{p} = \frac{1}{2} + \frac{1}{3} + \frac{1}{5} + \frac{1}{7} + \frac{1}{11} + \cdots$$

is divergent. This tells us something about how the infinitely many primes are sprinkled in among all the integers.

Bernoulli's Proof of the Divergence of the Harmonic Series

The usual way to show that the harmonic series diverges is to observe that when viewed in the form $1 + (\frac{1}{2}) + (\frac{1}{3} + \frac{1}{4}) + (\frac{1}{5} + \frac{1}{6} + \frac{1}{7} + \frac{1}{8}) + \cdots$, each grouped sum is at least $1/2$; hence the partial sum sequence increases without bound. An even shorter proof can be based on the comparison $H = 1 + \frac{1}{2} + \frac{1}{3} + \frac{1}{4} + \cdots > \frac{1}{2} + \frac{1}{2} + \frac{1}{4} + \frac{1}{4} + \cdots = 1 + \frac{1}{2} + \cdots = H$, an impossible inequality if H were finite.

An earlier argument communicated by Jakob Bernoulli in 1689, which he credits to his brother Johann, is very nice. He uses the familiar series $\sum_{n=M}^{\infty} \frac{1}{n(n+1)} = \frac{1}{M}$.* Following Bernoulli's own notation he writes

$$A = \frac{1}{2} + \frac{1}{3} + \frac{1}{4} + \frac{1}{5} + \frac{1}{6} + \cdots,$$

which is the same as

$$B = \frac{1}{2} + \frac{2}{6} + \frac{3}{12} + \frac{4}{20} + \frac{5}{30} + \cdots.$$

But B is the sum of all the following:

$$\begin{aligned}
C &= \tfrac{1}{2} + \tfrac{1}{6} + \tfrac{1}{12} + \tfrac{1}{20} + \cdots \\
D &= \phantom{\tfrac{1}{2} + {}} \tfrac{1}{6} + \tfrac{1}{12} + \tfrac{1}{20} + \cdots \\
E &= \phantom{\tfrac{1}{2} + \tfrac{1}{6} + {}} \tfrac{1}{12} + \tfrac{1}{20} + \cdots \\
F &= \phantom{\tfrac{1}{2} + \tfrac{1}{6} + \tfrac{1}{12} + {}} \tfrac{1}{20} + \cdots \\
\cdots &
\end{aligned}$$

and so $A = B = C + D + E + F + \cdots$
But if we add along diagonals indicated by the arrows, we see

$$\begin{aligned}
A = B = C + D + E + F + \cdots &= \left(\tfrac{1}{2} + \tfrac{1}{6} + \tfrac{1}{12} + \tfrac{1}{20} + \cdots\right) + \left(\tfrac{1}{6} + \tfrac{1}{12} + \tfrac{1}{20} + \cdots\right) \\
&\quad + \left(\tfrac{1}{12} + \tfrac{1}{20} + \cdots\right) + \left(\tfrac{1}{20} + \cdots\right) + \cdots \\
&= 1 + \tfrac{1}{2} + \tfrac{1}{3} + \tfrac{1}{4} + \cdots = 1 + A.
\end{aligned}$$

We now have $A = 1 + A$, which is impossible for a finite sum. (Question: If the series were convergent, could all the steps above be justified by the theorems you have learned about absolutely convergent series?)

A nice exposition of this can be found in "The Bernoullis and the Harmonic Series," by William Dunham in *The College Mathematics Journal*, volume 18 (January 1987), pages 18-23.

---

*This is an example of a telescoping series; note that $\frac{1}{M(M+1)} + \frac{1}{(M+1)(M+2)} + \cdots \frac{1}{(M+k)(M+k+1)} = (\frac{1}{M} - \frac{1}{M+1}) + (\frac{1}{M+1} - \frac{1}{M+2}) + \cdots + (\frac{1}{M+k} - \frac{1}{M+k+1}) = \frac{1}{M} + (-\frac{1}{M+1} + \frac{1}{M+1}) + (-\frac{1}{M+2} + \frac{1}{M+2}) + \cdots + (-\frac{1}{M+k} + \frac{1}{M+k}) - \frac{1}{M+k+1} = \frac{1}{M} - \frac{1}{M+k+1}$, which converges to $\frac{1}{M}$ as $k \to \infty$.

Partial Sums of the Harmonic Series

We know that the partial sums $h_n = 1 + \frac{1}{2} + \cdots + \frac{1}{n}$ of the harmonic series diverge to infinity at a very slow rate. Here we use the "third constant" $\gamma = 0.5772156649\cdots$ of Euler and the "second constant" e (also due to Euler!) to give an answer to this problem:

> Given any number $A > 0$, how many terms in the harmonic series will be sufficient to exceed A? That is, for what n_A are we assured $h_{n_A} > A$?

Let us recall* that the sequence $b_n = h_n - \ln(n + \frac{1}{2})$ decreases to γ, so that for all n we have $h_n - \ln(n + \frac{1}{2}) > \gamma$. Rewriting this inequality as $h_n > \gamma + \ln(n + \frac{1}{2})$, we see that we are assured $h_n > A$ once we have chosen n sufficiently large to have

(1) $$\gamma + \ln(n + \frac{1}{2}) \geq A.$$

Inequality (1) can be solved for n, giving

(2) $$n \geq e^{A-\gamma} - \frac{1}{2}.$$

To avoid taking a larger number of terms than necessary, we can take n_A to be the smallest integer that satisfies (2). By introducing the function $\lceil x \rceil$ = smallest integer $\geq x$, we can then take

(3) $$n_A = \lceil e^{A-\gamma} - \frac{1}{2} \rceil.$$

Example. For A=5, we find

$$n_5 = \lceil e^{5-\gamma} - \frac{1}{2} \rceil = \lceil 82.827\cdots \rceil = 83.$$

Indeed, $h_{82} = 4.9902\cdots$ and $h_{83} = 5.00206\cdots$, so in fact we needed at least 83 terms of the harmonic series to exceed the sum of 5.

Example. Since $n_{20} = \lceil e^{20-\gamma} - \frac{1}{2} \rceil = \lceil 272400599.55904078\cdots \rceil = 272400600$, this is the number of terms that guarantees us that $h_{n_{20}} > 20$. It can be shown that $h_{272400599} = 19.99999\ 99979\cdots$ and $h_{272400600} = 20.00000\ 00016\cdots$, and so we see that n_{20} is the least number of terms required in the harmonic series to exceed a sum of 20.

Example. Since $n_{100} = \lceil e^{100-\gamma} - \frac{1}{2} \rceil \approx 1.509 \times 10^{43}$, we see that approximately 1.509×10^{43} terms are required to reach a sum of 100. To appreciate this fully, suppose you write down the terms of the series $1 + \frac{1}{2} + \frac{1}{3} + \cdots$ at the rate of one per second, quitting when your sum reaches 100. This table will require 1.509×10^{43} seconds, or about 5×10^{33}

---

*See Euler's Constant, page 151.

centuries! Oh well, time flies when you're having fun.

Remark: $n_A = \lceil e^{A-\gamma} - \frac{1}{2} \rceil$ may, in some cases, be larger than needed for $h_{n_A} > A$. However, we will now show that $h_{n_A - 2} \le A$, and so h_{n_A} includes at most one unneeded term to exceed A.

In the section on Euler's Constant, it was shown that the sequence $h_n - \ln(n+1)$ increases monotonely to γ, which means that

$$h_n < \gamma + \ln(n+1). \tag{4}$$

Letting $n \le n_A - 2$, we then have from (4) that

$$h_n < \gamma + \ln(n_A - 1) < \gamma + \ln(e^{A-\gamma}) = A. \tag{5}$$

We have now proved the following result, due to L. Comtet:

Theorem: Let $A > 0$ and $n_A = \lceil e^{A-\gamma} - \frac{1}{2} \rceil$. Then the smallest integer n for which $h_n > A$ is either n_A or $n_A - 1$.

References: R. P. Boas, Jr., and J. W. Wrench, Jr., "Partial Sums of the Harmonic Series," *American Mathematics Monthly 78* (1971), 864-870.

L. Comtet, "Problems 5346," *American Mathematics Monthly 74* (1967), 209.

Evaluating the Sum of the Alternating Harmonic Series

Let's consider the alternating harmonic series

$$\sum_{n=1}^{\infty}(-1)^{n+1}\frac{1}{n} = 1 - \frac{1}{2} + \frac{1}{3} - \frac{1}{4} + \cdots .$$

The alternating series test tells us that the series converges to a sum S, but it does not tell us what S is. We know that S is the limit of the partial sum sequence $\{s_k\}$, and so S is also the sum of the subsequence $\{s_{2k}\}$, where

$$s_{2k} = 1 - \frac{1}{2} + \frac{1}{3} - \frac{1}{4} + \cdots + \frac{1}{2k-1} - \frac{1}{2k}.$$

Let us now relate s_{2k} to the partial sums

$$h_k = 1 + \frac{1}{2} + \cdots + \frac{1}{k}$$

of the harmonic series. To see the connection, we can proceed as follows:

$$\begin{aligned} s_{2k} &= 1 + \frac{1}{3} + \cdots + \frac{1}{2k-1} - \frac{1}{2} - \frac{1}{4} - \cdots - \frac{1}{2k} \\ &= 1 + \frac{1}{2} + \frac{1}{3} + \cdots + \frac{1}{2k-1} + \frac{1}{2k} - 2(\frac{1}{2} + \frac{1}{4} + \cdots + \frac{1}{2k}) \\ &= h_{2k} - h_k. \end{aligned}$$

We need to remember the connection that h_k has to $\ln k$, namely that the sequence $\{c_k\}$ defined by

$$c_k = h_k - \ln k$$

converges to the constant γ of Euler.* To use this fact, we first rewrite s_{2k} a bit:

$$s_{2k} = h_{2k} - h_k = [\ln(2k) + c_{2k}] - [\ln k + c_k] = \ln 2 + c_{2k} - c_k.$$

Now we can let $k \to \infty$ to evaluate S:

$$S = \lim_{k\to\infty} s_{2k} = \lim_{k\to\infty}[\ln 2 + c_{2k} - c_k] = \ln 2 + \gamma - \gamma = \ln 2.$$

In conclusion, we have shown that

$$1 - \frac{1}{2} + \frac{1}{3} - \frac{1}{4} + \frac{1}{5} - + \cdots = \ln 2.$$

---

*See Euler's Constant, page 151.

Evaluating Sums of Simple Rearrangements of the Alternating Harmonic Series

It has been shown that the alternating harmonic series is convergent and has sum $\ln 2$:

$$1-\frac{1}{2}+\frac{1}{3}-\frac{1}{4}+\frac{1}{5}-+\cdots=\ln 2.$$

What will happen if the order of the terms of the series is changed, a process that forms a *rearrangement* of the original series?

To be specific, let's consider the rearrangement

$$1+\frac{1}{3}+\frac{1}{5}-\frac{1}{2}-\frac{1}{4}+\frac{1}{7}+\frac{1}{9}+\frac{1}{11}-\frac{1}{6}-\frac{1}{8}+\frac{1}{13}+\cdots.$$

Here we take 3 positive, 2 negative, 3 positive, 2 negative terms, and so on. Since the order of the positive terms and the order of the negative terms is not changed, this is an example of a *simple* rearrangement.

We will now want to see if the rearranged series is still convergent and, if so, what its sum is. Thus we must examine the sequence of partial sums: $s_1=1, s_2=1+\frac{1}{3}, s_3=1+\frac{1}{3}+\frac{1}{5}, s_4=1+\frac{1}{3}+\frac{1}{5}-\frac{1}{2},\cdots$.

Let's concentrate on the partial sums that contain full blocks of the 3 positive terms followed by the 2 negative terms. We see that

$$s_5 = 1+\frac{1}{3}+\frac{1}{5}-\frac{1}{2}-\frac{1}{4}$$

$$s_{5\cdot 2} = 1+\frac{1}{3}+\frac{1}{5}-\frac{1}{2}-\frac{1}{4}+\frac{1}{7}+\frac{1}{9}+\frac{1}{11}-\frac{1}{6}-\frac{1}{8}$$

$$s_{5m} = 1+\frac{1}{3}+\cdots+\frac{1}{2(3m)-5}+\frac{1}{2(3m)-3}+\frac{1}{2(3m)-1}$$

$$-\frac{1}{2(2m)-2}-\frac{1}{2(2m)}.$$

Since the partial sums involve just finitely many terms, their values are not changed by moving their terms about. As an example, let's juggle $s_{5\cdot 2}$ into a new form by adding and subtracting $\frac{1}{2}+\frac{1}{4}+\cdots+\frac{1}{12}$.

$$\begin{aligned}
s_{5\cdot 2} &= 1+\tfrac{1}{3}+\tfrac{1}{5}-\tfrac{1}{2}-\tfrac{1}{4}+\tfrac{1}{7}+\tfrac{1}{9}+\tfrac{1}{11}-\tfrac{1}{6}-\tfrac{1}{8}\\
&= \left(1+\tfrac{1}{2}+\tfrac{1}{3}+\tfrac{1}{4}+\tfrac{1}{5}+\tfrac{1}{6}+\tfrac{1}{7}+\tfrac{1}{8}+\tfrac{1}{9}+\tfrac{1}{10}+\tfrac{1}{11}+\tfrac{1}{12}\right)-\tfrac{1}{2}-\tfrac{1}{4}-\tfrac{1}{6}-\tfrac{1}{8}\\
&\quad -\left(\tfrac{1}{2}+\tfrac{1}{4}+\tfrac{1}{6}+\tfrac{1}{8}+\tfrac{1}{10}+\tfrac{1}{12}\right)\\
&= \left(1+\tfrac{1}{2}+\tfrac{1}{3}+\cdots+\tfrac{1}{12}\right)-\tfrac{1}{2}\left(1+\tfrac{1}{2}+\tfrac{1}{3}+\tfrac{1}{4}\right)\\
&\quad -\tfrac{1}{2}\left(1+\tfrac{1}{2}+\cdots+\tfrac{1}{6}\right).
\end{aligned}$$

This suggests introducing the partial sums

$$h_n = 1 + \frac{1}{2} + \frac{1}{3} + \cdots + \frac{1}{n}$$

of the harmonic series $\sum \frac{1}{n}$, since then we have

$$s_{5\cdot 2} = h_{2\cdot 2\cdot 3} - \frac{1}{2}h_{2\cdot 2} - \frac{1}{2}h_{2\cdot 3}.$$

The same algebraic sleight of hand applied to s_{5m} shows that

$$s_{5m} = h_{2\cdot 3m} - \frac{1}{2}h_{2m} - \frac{1}{2}h_{3m}.$$

Next let us recall* that the sequence c_k, defined by

$$c_k = h_k - \ln k,$$

is convergent to the Euler constant γ. Thus we can write s_{5m} in the form

$$s_{5m} = [\ln(2\cdot 3\cdot m) + c_{2\cdot 3\cdot m}] - \frac{1}{2}[\ln(2m) + c_{2m}] - \frac{1}{2}[\ln(3m) + c_{3m}].$$

Using $\ln(2\cdot 3\cdot m) = \ln 2 + \ln 3 + \ln m$ and similar expansions for the other two log terms, we get

$$\begin{aligned} s_{5m} &= \ln 2 + \ln 3 + \ln m + c_{2\cdot 3m} \\ &-\frac{1}{2}\ln m - \frac{1}{2}\ln 2 - \frac{1}{2}c_{2m} \\ &-\frac{1}{2}\ln m - \frac{1}{2}\ln 3 - \frac{1}{2}c_{3m} \\ &= \ln 2 + \frac{1}{2}\ln\frac{3}{2} + c_{2\cdot 3m} - \frac{1}{2}c_{2m} - \frac{1}{2}c_{3m}. \end{aligned}$$

We can now take the limit:

$$\begin{aligned} \lim_{m\to\infty} s_{5m} &= \ln 2 + \tfrac{1}{2}\ln\tfrac{3}{2} + \gamma - \tfrac{1}{2}\gamma - \tfrac{1}{2}\gamma \\ &= \ln 2 + \tfrac{1}{2}\ln\tfrac{3}{2}, \end{aligned}$$

which shows that the subsequence s_{5m} of the partial sum sequence converges to $\ln 2 + \frac{1}{2}\ln\frac{3}{2}$.

In order to deal with the full sequence s_k, we write each integer k in quotient plus remainder form,

$$k = 5m + r, \quad 0 \le r < 5.$$

---

*See Euler's Constant, page 151.

Thus m is the integer part of $\frac{k}{5}$. Since s_{5m} includes all the terms through the one of magnitude $\frac{1}{2 \cdot 2m}$, we then have that

$$s_k = s_{5m} + \quad (r \text{ more terms, each in absolute value smaller than } \frac{1}{2 \cdot 2m}).$$

As $k \to \infty$, we also have $m \to \infty$, and so

$$\lim_{k \to \infty} s_k = \lim_{m \to \infty} s_{5m} + 0 = \ln 2 + \frac{1}{2} \ln \frac{3}{2}.$$

We have now shown that rearranging the alternating harmonic series into blocks of 3 positive and 2 negative terms is a convergent series, with sum $\ln 2 + \frac{1}{2} \ln \frac{3}{2}$.

The very same procedure leads to a general result:

A simple rearrangement of the alternating harmonic series into blocks of p positive and n negative terms has the sum $\ln 2 + \frac{1}{2} \ln \frac{p}{n}$.

The key algebraic step is to show that the partial sum with m blocks of $p + n$ terms has the form

$$(*) \quad s_{m(p+n)} = h_{2mp} - \frac{1}{2} h_{mn} - \frac{1}{2} h_{mp}.$$

Proceeding as before, this leads to

$$(**) \quad s_{(p+n)m} = \ln 2 + \frac{1}{2} \ln \frac{p}{n} + c_{2mp} - \frac{1}{2} c_{mn} - \frac{1}{2} c_{mp}.$$

Exercises on Evaluating Simple Rearrangements of the Alternating Harmonic Series

1. Verify (*), page 179.

2. Verify (**), page 179.

3. Find (that is, give values of p and n) simple rearrangements of the alternating harmonic series that

   (a) sum to $\ln 2 + \frac{1}{2} \ln \frac{7}{13}$

   (b) sum to 0

   (c) exceed the sum 10

   (d) are less than -20

Answers to Exercises on Evaluating Simple Rearrangements of the Alternating Harmonic Series

1.
$$s_{m(p+n)} = \sum_{r=0}^{m-1}\left[\sum_{j=1}^{p}\frac{1}{2rp+2j-1} - \sum_{k=1}^{n}\frac{1}{2rn+2k}\right]$$

$$= \sum_{r=0}^{m-1}\left[\sum_{i=1}^{2p}\frac{1}{2rp+i} - \sum_{i=1}^{p}\frac{1}{2rp+2i} - \sum_{k=1}^{n}\frac{1}{2rn+2k}\right]$$

$$= h_{2mp} - \frac{1}{2}h_{mp} - \frac{1}{2}h_{mn}$$

2.
$$s_{m(p+n)} = [\ln(2mp) + c_{2mp}] - \frac{1}{2}[\ln(mp) + c_{mp}] - \frac{1}{2}[\ln(mn) + c_{mn}]$$

$$= \ln 2 + \frac{1}{2}\ln\frac{mp}{mn} + c_{2mp} - \frac{1}{2}c_{mp} - \frac{1}{2}c_{mn}.$$

3. (a) $p = 7, n = 13$, or $p = 14, n = 26$ just so $\frac{p}{n} = \frac{7}{13}$.

(b) $\frac{p}{n} = \frac{1}{4}$ For example, $p = 1$ and $n = 4$, since then $\ln 2 + \frac{1}{2}\ln\frac{1}{4} = 0$.

(c) You need $\ln 2 + \frac{1}{2}\ln\frac{p}{n} > 10$; equivalently, $\frac{p}{n} > \frac{1}{4}e^{10} \approx 5506.6$.

Thus $p = 5507$ and $n = 1$ will do it.

(d) You need $\ln 2 + \frac{1}{2}\ln\frac{p}{n} < -20$; equivalently,

$$4\frac{p}{n} < e^{-40}, \text{ or } \frac{n}{p} > 4e^{40}.$$

Thus, as a calculator will show in scientific notation, $p = 1$ and $n = 9.5 \times 10^{17}$ will suffice. (Is the size of the negative blocks about what you would have guessed? If so, you've got quite an intuition!)

The Harmonic Series and a Stack of Cards

How far can a card two units long hang over the edge of a table and not fall? The obvious answer: one unit.

Let us show that cards arranged in each of the following figures will not fall off the table.

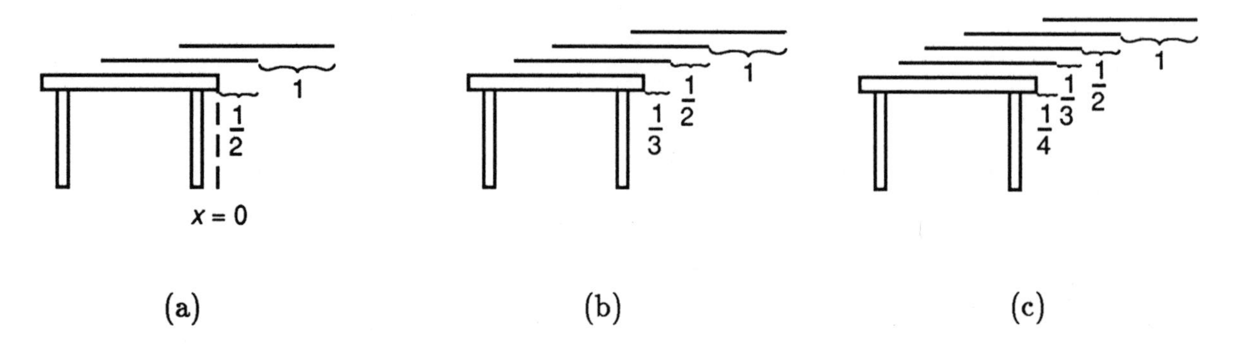

Note for (a), the moment about $x = 0$ for the two cards is $-\frac{1}{2}(1) + \frac{1}{2}(1) = 0$, so the cards balance. Also, the top card doesn't fall off the first one. In (b), the top two cards don't fall off the bottom one, that is, just case (a). The three have moment $(-\frac{2}{3})(1) - \frac{1}{6}(1) + \frac{5}{6}(1) = 0$ about the edge of the table, so they balance.

Similarly, it can be shown that cards (each of length 2 units) arranged as in (d) do not fall.

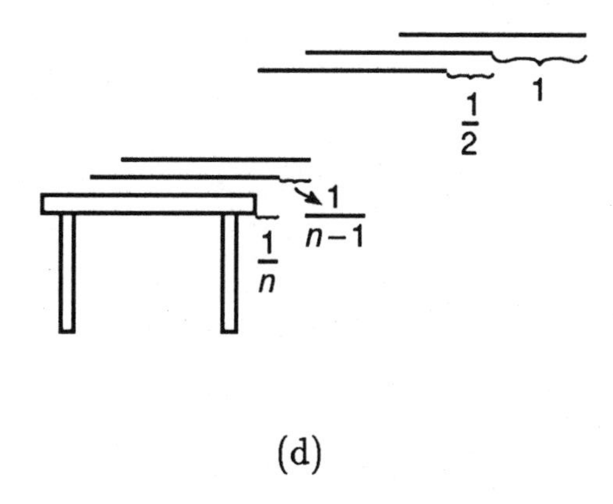

(d)

Exercises on the Harmonic Series and a Stack of Cards

1. If one has a limitless supply of cards and no glue, how far can they be extended past the edge of the table?

2. Suppose you want to support a pebble that weighs the same as one card at the edge of the last card in a stack. How could the cards be arranged to do this? What kind of configuration of cards would support a pebble three times as heavy? How far out can you build a bridge of overhanging cards (so you could stand on it as you would a diving board)? Use ordinary playing cards and no glue!

Answers to Exercises on the Harmonic Series and a Stack of Cards

1. As far as you wish, since $\sum \frac{1}{n}$ diverges.

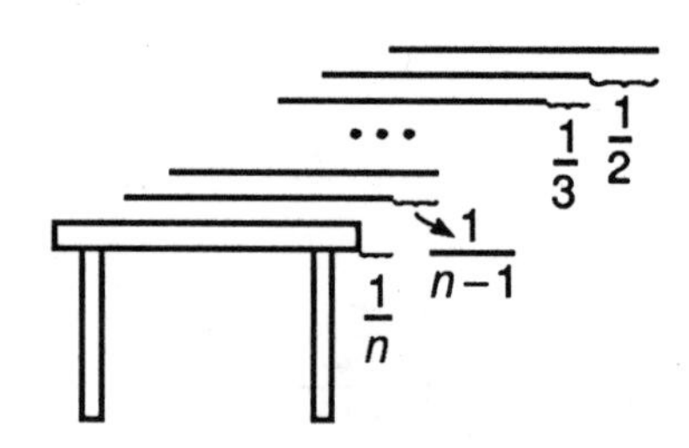

2. The configuration shown above will support a pebble of one unit mass at its edge, since it supported the last card in figure (d). All we have done is replace the top card with a pebble of equal mass. Similarly, the top 3 cards can be removed (last overhang is 1/4) and a pebble of mass 3 placed on the edge of the last card. Continuing this reasoning, sum the series $1+\frac{1}{2}+\cdots+\frac{1}{n}$ until this sum exceeds your mass. Remove N cards from the stack arranged as in (d) and you can walk to the edge of the remaining ones without falling. By using enough cards even after the top N are removed, the remaining overhang can be made as large as we wish!

   If you only wish to extend past the table one yard, it will take approximately 482,365,609 cards. If each card is 0.01 inches thick, this requires a stack of cards 76 miles high. An ordinary deck of 52 playing cards 3 1/2 inches long would extend about 8 inches past the edge, since $(1\frac{3}{4}) \times (1+\frac{1}{2}+\frac{1}{3}+\frac{1}{4}+\frac{1}{5}+\cdots+\frac{1}{52}) = (1.75)(4.538043951) = 7.941576915$.

More on a Stack of Cards

Suppose n cards, each two units long, are placed with overhangs of length $x_1, x_2, \ldots, x_n$ as shown. Can they be placed so they would extend further than they do when we make $x_1 = 1, x_2 = \frac{1}{2}, \ldots, x_n = \frac{1}{n}$, as we did before?* Computing centroids in a routine way, the conditions that no part of the stack will fall and that no cards recede are the following:

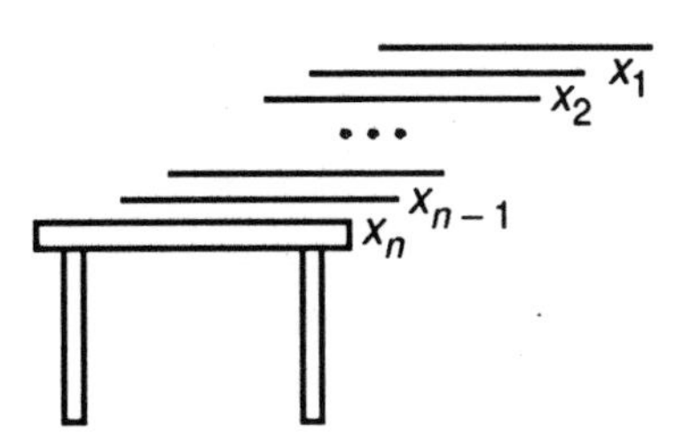

(1.) $x_i \geq 0$ for $i = 1, 2, 3 \ldots, n$ (no cards recede)

(2.) $x_1 \leq 1$ (the top card must not fall off the second one)

(3.) $x_2 + \frac{1}{2}x_1 \leq 1$ (the top two cards must not fall off the third one)

(4.) $x_3 + \frac{2}{3}x_2 + \frac{1}{3}x_1 \leq 1$ (the top three cards must not fall off the fourth one)

$\vdots$

$x_n + \frac{n-1}{n}x_{n-1} + \cdots + \frac{1}{n}x_1 \leq 1$ (the whole stack must stay on the table)

Our problem is to maximize the quantity $x_1 + x_2 + \cdots + x_n$ subject to all $n+1$ of the constraints above. This is a different kind of max-min problem than we have worked with calculus tools. There is an entire field in mathematics, developed during and since World War II, called Operations Research (OR for short), which deals with problems like this one. Applying these tools, it can be shown that the harmonic lengths for the overhangs are optimal.

The constraints are relaxed if you are allowed to glue the cards together. Note in the figure below that the moment about $x = 0$ for all three cards is $(-1)\times 1 + 0 \times 1 + 1 \times 1 = 0$; yet the stack falls (try it.) In this case we have $x_1 = x_2 = 1$ and $x_3 = 0$, so constraint number (3.) is violated and the top two cards fall off the third. If the stack is glued together (still assuming $x_i \geq 0$), only the last constraint is needed because a partial stack cannot fall.

*See The Harmonic Series and a Stack of Cards, page 182.

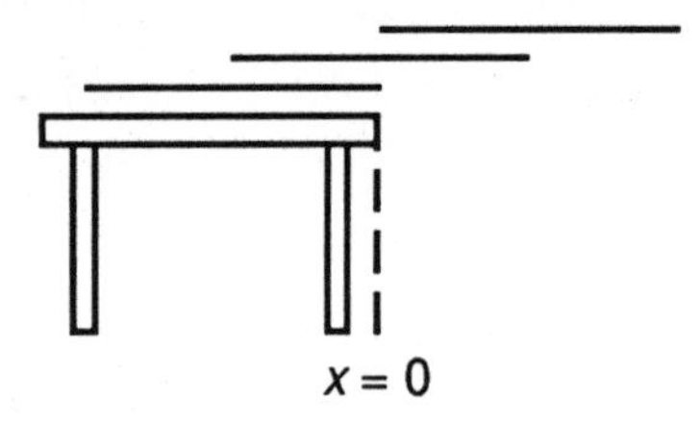

Note also that the overhang of the three glued cards is two units past the end of the table, whereas the best we could do with three unglued cards was $1 + \frac{1}{2} + \frac{1}{3} = 1\frac{5}{6}$ units. It can be shown that $x_3 = 0, x_1 = x_2 = 1$ is optimal for three glued cards under constraints (1) and (4). (If constraint (1) is dropped and, using glue, only constraint (4) applies, the overhang of the three cards can be made longer than two units–do you see how?)

This illustrates how changing the physical characteristic of a problem changes its mathematical description and that tools beyond those we have learned so far will be required for certain types of max-min problems.

Remark: If one allows more than one card (or trick) per layer, there are "nonharmonic" stacking arrangements that achieve a larger overhang for a given number of cards. See J.E. Drummond, "On Stacking Tricks to Achieve a Large Overhang," *Mathematics Gazette* **65** (March, 1981), 40-42.

The Jellybean Problem–More on the Harmonic Series

Consider the following problem: Carol does not like red-cherry jellybeans, but has r of them mixed in with her bag of g green-lime jellybeans. Every day Carol reaches in her bag and takes a jellybean at random. If it is green she eats it, but if it is red she puts it back in the bag. What is the expected number of days until Carol consumes the last green jellybean?"

Let $E(r, g)$ denote the expected number of days to consume the last green jellybean, where r and g are the respective number of red and green jellybeans. Let us assume for the moment only one green jellybean is in the bag, so $g = 1$. If a red jellybean is drawn on the first day, which has probability $\frac{r}{1+r}$, Carol's expected number of days to draw the green one is extended by one day. If she is lucky enough to draw a green, which has probability $\frac{1}{r+1}$, she is done. The following equation expresses this information:

$$E(r, 1) = \frac{r}{1+r}[E(r, 1) + 1] + \frac{1}{1+r}.$$

Solving for $E(r, 1)$ we find

$$E(r, 1) = 1 + r. \tag{1}$$

Next, suppose there are $g \geq 2$ green jellybeans. If Carol draws a red on day one, with probability $\frac{r}{r+g}$, she knows she has extended her expected time to eat the last green one by a day. If Carol draws and eats one of the green jellybeans, with probability $\frac{g}{r+g}$, this takes one day and leaves her with $E(r, g - 1)$ expected days to draw the last of the remaining $g - 1$ green jellybeans. The relationship just described in words is expressed by the equation

$$E(r, g) = \frac{r}{r+g}[E(r, g) + 1] + \frac{g}{r+g}[E(r, g - 1) + 1].$$

By algebraic rearrangement, this equation has the simpler form

$$E(r, g) = E(r, g - 1) + 1 + \frac{r}{g}, \quad g \geq 2. \tag{2}$$

Taking $g = 2$, $E(r, 2) = E(r, 1) + 1 + \frac{r}{2}$, and so using equation (1) we find $E(r, 2) = 2 + r(1 + \frac{1}{2})$. Indeed, from (1) and (2) it follows by mathematical induction that

$$E(r, g) = g + r(1 + \frac{1}{2} + \cdots + \frac{1}{g}). \tag{3}$$

We note the appearance of $h_g = 1 + \frac{1}{2} + \cdots + \frac{1}{g}$, the gth partial sum of the harmonic series $1 + \frac{1}{2} + \frac{1}{3} + \cdots$.

Numerical Examples

A. If $r = 5$ red-cherry and $g = 20$ green-lime jellybeans, the expected number of days to consume the last green jellybean is

$$\begin{aligned} E(5,20) &= 20 + 5(1 + \frac{1}{2} + \cdots + \frac{1}{20}) \\ &\approx 20 + 5(3.597) \\ &= 37.989. \end{aligned}$$

B. If $r = 10$ red-cherry and $g = 40$ green-lime jellybeans, the expected number of days to consume the last green jellybean is

$$\begin{aligned} E(10,40) &= 40 + 10(1 + \frac{1}{2} + \cdots + \frac{1}{40}) \\ &\approx 40 + 10(4.27854) \\ &= 82.7854. \end{aligned}$$

Exercises on the Jellybean Problem

1. Jane has one red-cherry and ten green-lime jellybeans and Kate has ten red-cherry and one green-lime jellybean. Their taste in jellybeans is exactly like Carol's. Who, Jane or Kate, do you expect will take the longest time to get to the last green jellybean?

2. (See Euler's Constant, p. 151.) Recall that $h_n \approx \ln(n+\frac{1}{2})+\gamma$, where $\gamma = 0.57721\cdots$ is Euler's constant. Use this approximation to estimate the expected number of days Carol requires if her bag initially contains 200 jellybeans, 100 of each kind.

Answers to Exercises on the Jellybean Problem

1. Jane: $E(1,10) = 10 + 1(1 + \frac{1}{2} + \cdots + \frac{1}{10}) \approx 10 + 2.929 = 12.929$
   Kate: $E(10,1) = 1 + 10(1) = 11.$

   Jane is therefore expected to take nearly two days longer than Kate.

2. $$\begin{aligned} E(100,100) &= 100 + 100(1 + \tfrac{1}{2} + \cdots + \tfrac{1}{100}) \\ &\approx 100 + 100[\ln(100.5) + 0.57721] \\ &\approx 618.7 \text{ days} \end{aligned}$$

   [that is , about one year and $8\frac{1}{3}$ months!]

Constructing the Tangent Lines to Conics

Given a point P on a circle, it is a simple matter to construct the tangent to the circle at P; just take the line which is perpendicular to the radius through P. Differential calculus allows one to discover ways to construct the tangent lines to any conic section, as the following exercises reveal.

1. Let A and B be the x and y intercepts of the line tangent to the hyperbola $y = k/x$ at $P = (x_0, y_0)$. Let $C = (x_0, 0)$ and $D = (0, y_0)$. Show:

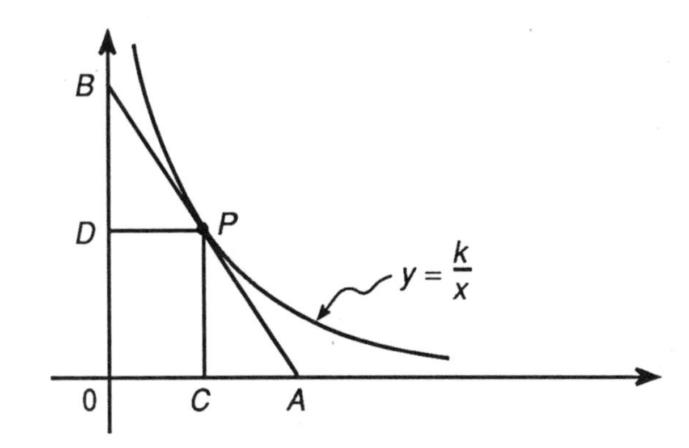

(a) $\overline{CD} \| \overline{AB}$

(b) $P = \text{midpoint } \overline{AB}$

(c) area $\triangle OAB = 2k$

(d) Explain how properties (a) and (b) can be used to draw the tangent line to $y = 8/x$ at $P(2, 4)$.

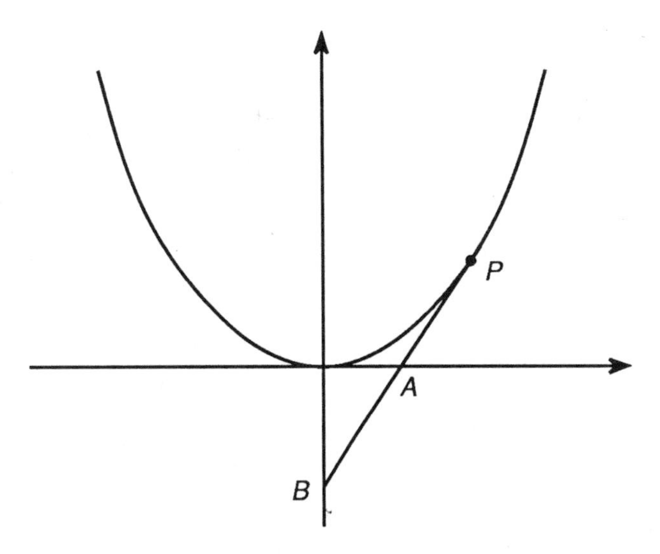

2. Let A and B be the x and y intercepts of the line tangent to the parabola $y = cx^2$ at $P = (x_0, y_0)$, where $y_0 > 0$.

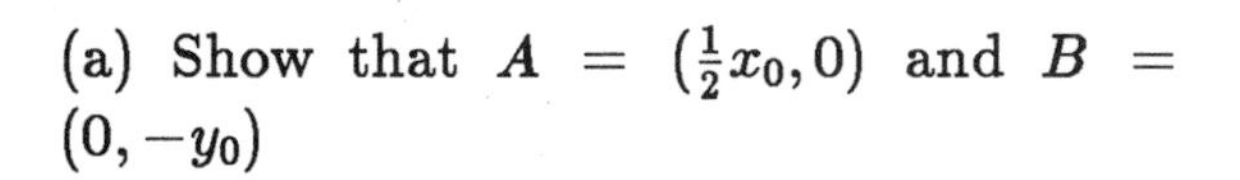

(a) Show that $A = (\frac{1}{2}x_0, 0)$ and $B = (0, -y_0)$

(b) Use part (a) to construct the tangent line to $y = \frac{1}{2}x^2$ at $P = (4, 8)$.

3. Let $X = (a, 0)$ and $Y = (0, b)$ be points of the positive x and y axes, and suppose a ray $\overrightarrow{OR}$ at angle φ to the x axis intersects the circles of radius a and b at points $C = (a\cos\varphi, a\sin\varphi)$ and $D = (b\cos\varphi, b\sin\varphi)$. Let $P = (a\cos\varphi, b\sin\varphi)$.

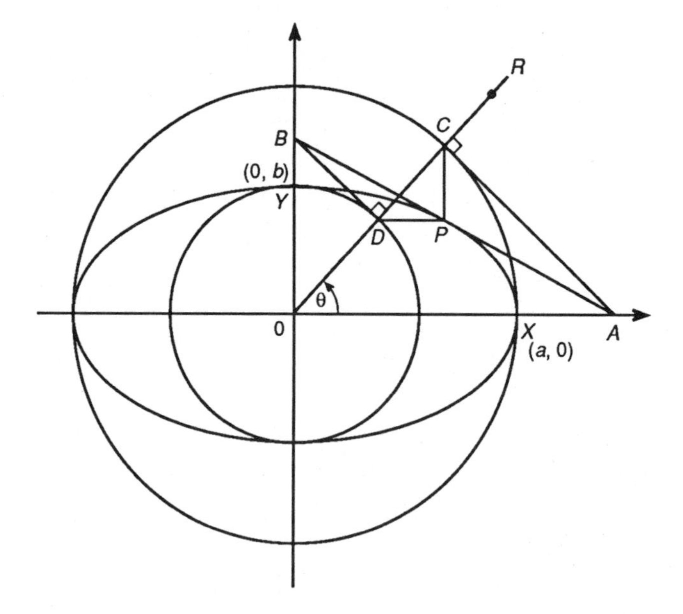

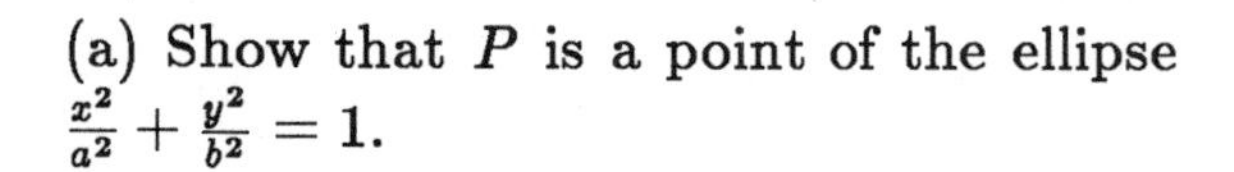

(a) Show that P is a point of the ellipse $\frac{x^2}{a^2} + \frac{y^2}{b^2} = 1$.

(b) Show that the line tangent to the ellipse at $P = (x_0, y_0)$ in the first quadrant has slope $-\frac{x_0 b^2}{y_0 a^2}$, equation $\frac{x x_0}{a^2} + \frac{y y_0}{b^2} = 1$, and intercepts the x and y axes at $A = (\frac{a^2}{x_0}, 0)$, $B = (0, \frac{b^2}{y_0})$.

(c) Show that $\overline{AC}$ and $\overline{BD}$ are both perpendicular to the ray $\overrightarrow{OR}$.

(d) Use the information in (a), (b), and (c) to construct the line tangent to the ellipse $\frac{x^2}{9} + \frac{y^2}{4} = 1$ at the point $P = (\frac{3}{2}, \sqrt{3})$.

Answers to Exercises on Constructing the Tangents to Conics

1. (a) $y'_0 = -\frac{c}{x_0^2} = -\frac{c}{x_0} \cdot \frac{1}{x_0} = -\frac{y_0}{x_0}$, which is the slope of the line from $D = (0, y_0)$ to $C = (x_0, 0)$.

   (b) The x-axis and the horizontal line $\overleftrightarrow{DP}$ intercept equal segments of the parallel lines $\overleftrightarrow{CD}$ and $\overleftrightarrow{AB}$, so $CD = AP$. Similarly $CD = PB$, and so $AP = PB$.

   (c) Since $OA = 2OC = 2x_0$ and $OB = 2OD = 2y_0$, we see area $\triangle OAB = \frac{1}{2}(2x_0)(2y_0) = 2x_0y_0 = 2x_0(\frac{k}{x_0}) = 2k$. Thus the tangent lines to $y = k/x$ cut off triangles of constant area $2k$.

   (d) Locate $A = (2x_0, 0) = (4, 0)$ and $B = (0, 2y_0) = (0, 8)$. The line segment $\overline{AB}$ is tangent to the hyperbola at its midpoint $P = (2, 4)$.

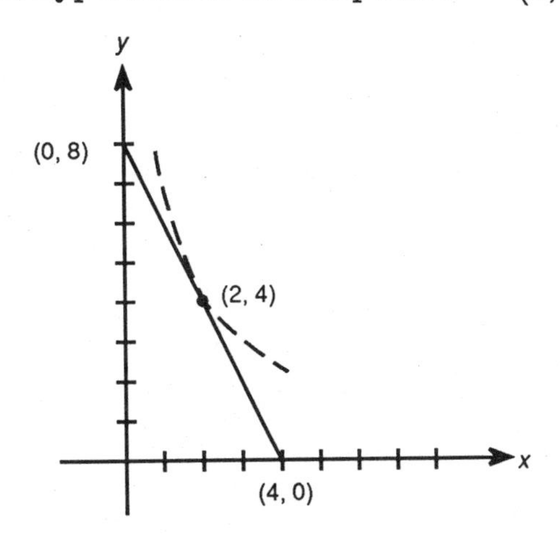

2. (a) $y'_0 = 2cx_0 = \frac{2cx_0^2}{x_0} = \frac{2y_0}{x_0}$ is the slope of the tangent line at $P = (x_0, y_0)$. Using the point-slope equation, the tangent line has equation $y - y_0 = \frac{2y_0}{x_0}(x - x_0)$; that is, $y = \frac{2y_0}{x_0}x - y_0$. Setting $y = 0$, the x-intercept, x_1, is seen to be $x_1 = \frac{1}{2}x_0$. Similarly, setting $x = 0$ shows the y-intercept is $y_1 = -y_0$.

   (b) To construct the tangent line to $y = \frac{1}{2}x^2$ at $P = (4, 8)$, locate $A = (\frac{x_0}{2}, 0) = (2, 0)$ and $B = (0, -y_0) = (0, -8)$. Then $\overleftrightarrow{AB}$ is the line tangent to the parabola at $P = (4, 8)$.

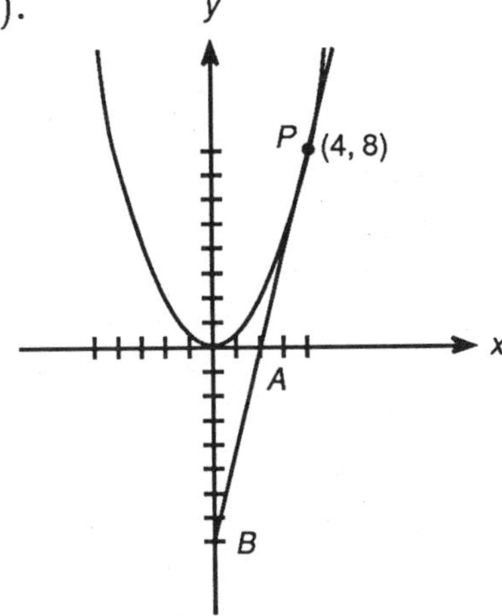

3. (a) If $P = (x_0, y_0) = (a\cos\varphi, b\sin\varphi)$, then $\frac{x_0^2}{a^2} + \frac{y_0^2}{b^2} = \cos^2\varphi + \sin^2\varphi = 1$.

(b) $y = b\sqrt{1 - \frac{x^2}{a^2}}$, $y_0' = \frac{b(-x_0/a^2)}{\sqrt{1-x_0^2/a^2}} = \frac{-b^2x_0/a^2}{b\sqrt{1-x_0^2/a^2}} = \frac{-b^2x_0}{a^2y_0}$. Since the line with equation $\frac{xx_0}{a^2} + \frac{yy_0}{b^2} = 1$ has slope y_0' and contains $P_0 = (x_0, y_0)$, it is necessarily the equation of the tangent line. Setting $y = 0$ and $x = 0$, the axis intercepts of the tangent line are $x_1 = a^2/x_0$, $y_1 = b^2/y_0$.

(c) Let $\hat{A}$ be the point on the x-axis for which $\overrightarrow{\hat{A}C} \perp \overrightarrow{OR}$:

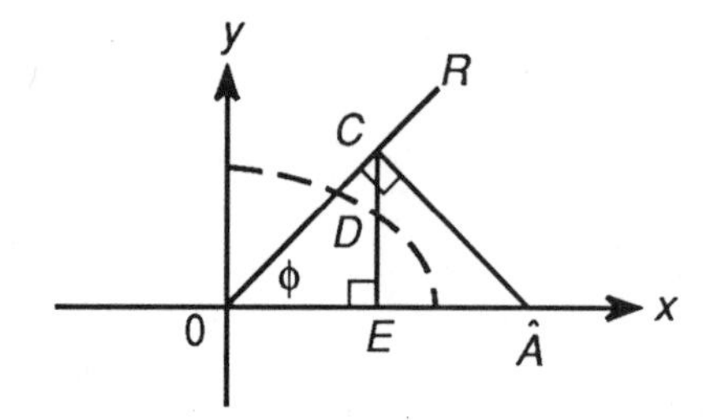

By similar right triangles, $\triangle OEC \sim \triangle OC\hat{A}$, so we have $OC : OE = O\hat{A} : OC$ or $O\hat{A} = (OC)^2/OE = a^2/x_0 = OA$. Hence $\hat{A} = A$, and $\overline{AC} \perp \overrightarrow{OR}$. An analogous argument shows $\overline{BD} \perp \overrightarrow{OR}$.

(d) We have $a = 3, b = 2, x_0 = 3/2$, and $y_0 = \sqrt{3}$. Thus $a^2/x_0 = 6$ and $b^2y_0 = 4/\sqrt{3} \approx 2.3$. Locating $A = (6,0)$ and $B = (0, 4/\sqrt{3})$, we know the segment $\overline{AB}$ will be tangent to the ellipse at $P = (3/2, \sqrt{3})$.

Exercises on Obtaining Tangent Line Equations to General Conics

There is an easy routine to give the equation of a line tangent to the curve $Ax^2 + Bxy + Cy^2 + Dx + Ey + F = 0$ at an arbitrary point (x_0, y_0) on the curve.

1. The locus of points satisfying $4x^2 + 4xy + y^2 + 15x - 3y + 21 = 0$ is the parabola shown.

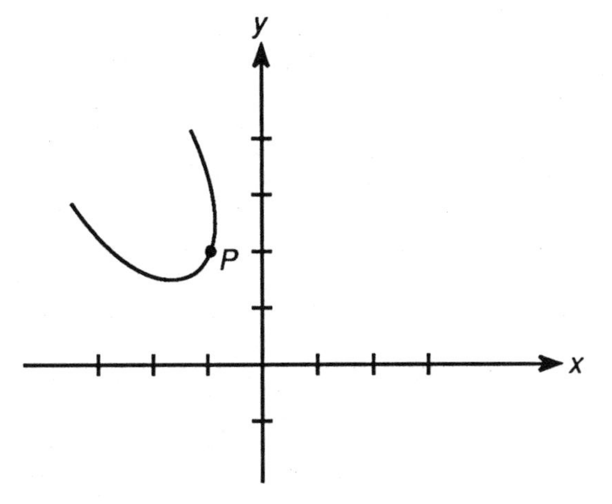

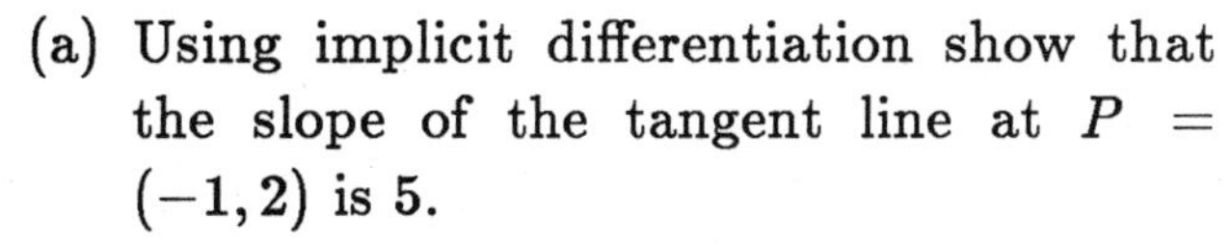

   (a) Using implicit differentiation show that the slope of the tangent line at $P = (-1, 2)$ is 5.

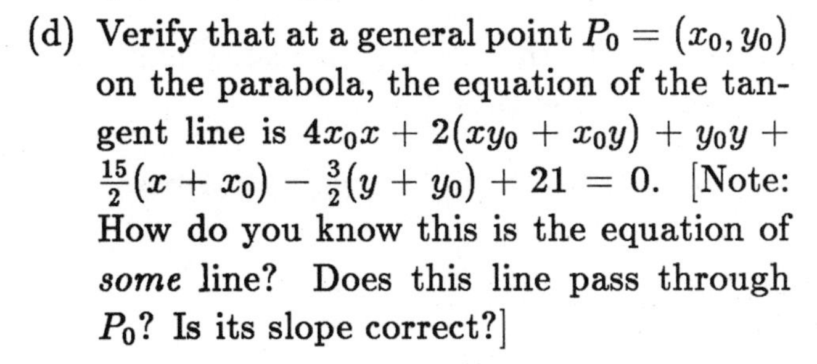

   (b) Show that the equation of the tangent line at P is $(*)$ $5x - y + 7 = 0$.

   (c) Show the equation $(*)$ can be written in the form $4(-1)x + 2[x(2) + (-1)y] + 2y + \frac{15}{2}(x - 1) - \frac{3}{2}(y + 2) + 21 = 0$.

   (d) Verify that at a general point $P_0 = (x_0, y_0)$ on the parabola, the equation of the tangent line is $4x_0x + 2(xy_0 + x_0y) + y_0y + \frac{15}{2}(x + x_0) - \frac{3}{2}(y + y_0) + 21 = 0$. [Note: How do you know this is the equation of *some* line? Does this line pass through P_0? Is its slope correct?]

2. The general quadratic equation (Q) $Ax^2 + Bxy + Cy^2 + Dx + Ey + F = 0$ represents one of the conic sections. If $P_0 = (x_0, y_0)$ is a point on the curve, show that the equation of the line tangent to the curve at P_0 is

   (T)
$$Axx_0 + \frac{1}{2}B(xy_0 + x_0y) + Cyy_0$$
$$+\frac{1}{2}D(x + x_0) + \frac{1}{2}E(y + y_0) + F = 0.$$

   Note that Ax^2 in (Q) corresponds to Axx_0 in (T). Describe how the other terms in (T) are obtained from corresonding terms in (Q).

3. Find the equation of the tangent line to the ellipse with equation

$$3x^2 + 4xy + 2y^2 + 3x + y - 11 = 0$$

   at $(2, -1)$.

Answers to Exercises on Obtaining Tangent Line Equations to General Conics

1. By implicit differentiation, evaluating at $x = x_0$ and $y = y_0$, we get

$$8x_0 + 4x_0y_0' + 4y_0 + 2y_0y_0' + 15 - 3y_0' = 0.$$

Dividing by 2 and rearranging terms we get

$$(**) \qquad (4x_0 + 2y_0 + \frac{15}{2}) + (2x_0 + y_0 - \frac{3}{2})y_0' = 0.$$

(a) At P we see $y_0' = -\dfrac{4(-1)+2(2)+15/2}{2(-1)+(2)-3/2} = 5.$

(b) $y - 2 = 5(x+1)$, or $5x - y + 7 = 0$.

(c) Simple algebra.

(d) The equation given can be written as

$$(4x_0 + 2y_0 + \frac{15}{2})x + (2x_0 + y_0 - \frac{3}{2})y + \frac{15}{2}x_0 - \frac{3}{2}y_0 + 21 = 0.$$

Comparing to equation (**) shows that the slope is correct, including the case of a vertical tangent [that is, when $2x_0 + y_0 - \frac{3}{2} = 0$].

2. Implicitly differentiating (Q), dividing by 2, and evaluating at P_0, we see

$$Ax_0 + \frac{B}{2}(x_0y_0' + y_0) + Cy_0y_0' + \frac{D}{2} + \frac{E}{2}y_0' = 0;$$

$$\text{that is, } (Ax_0 + \frac{B}{2}y_0 + \frac{D}{2}) + (\frac{B}{2}x_0 + Cy_0 + \frac{E}{2})y_0' = 0.$$

Rewriting (T) as

$$(Ax_0 + \frac{3}{2}y_0 + \frac{D}{2})x + (\frac{B}{2}x_0 + Cy_0 + \frac{E}{2})y$$
$$+\frac{1}{2}Dx_0 + \frac{1}{2}Ey_0 + F = 0$$

shows (T) has the correct slope. It clearly passes through P_0. The transformation rules are: $Ax^2 \longmapsto Axx_0, Bxy \longmapsto \frac{1}{2}B(xy_0 + x_0y), Cy^2 \longmapsto Cyy_0, Dx \longmapsto \frac{1}{2}D(x + x_0), Ey \longmapsto \frac{1}{2}E(y + y_0), F \longmapsto F.$

3. $3x^2 + 4xy + 2y^2 + 3x + y - 11 = 0$ at $P_0(2, -1)$ gives to the tangent equation

$$3x(2) + \frac{1}{2}\cdot 4(x(-1) + (2)y) + 2y(-1) + \frac{1}{2}\cdot 3(x+2) + \frac{1}{2}\cdot 1(y-1) - 11 = 0$$

or

$$(6 - 2 + \frac{3}{2})x + (4 - 2 + \frac{1}{2})y + (3 - \frac{1}{2} - 11) = 0.$$

This simplifies to $11x + 5y - 17 = 0$.

Billiard Balls and Conic Sections

The shapes of the conic sections give them distinctive reflective properties that only they have. These are interesting in themselves, but they may be put to practical use. It is said that Archimedes proposed that the reflective properties of the parabola be used to focus sunlight on enemy ships in the harbor of Syracuse in an attempt to ignite them. This application has its modern counterpart in a 31-foot-diameter solar furnace that generates temperatures up to 5400° F at the focus of the parabola.

Reflective properties of many figures have been studied and there is nice literature on "mathematical billiards." During the moon landing of Apollo 11, a 46 x 46 cm panel of 100 fused silica corner cubes (what you see in the corner of a room where three planes meet at right angles) each 3.8 cm in diameter, was left on the moon. A laser beam entering a corner cube bounces off all three planes and leaves in the direction parallel to the entering beam. (This experiment is easily duplicated by throwing a rubber ball into the corner of a room so that the ball hits all three planes.) Laser beams originating on earth and reflected off that panel are received back on earth and have been used to get very precise measures on earth-moon distances.

Let us now see what these billiard properties of the conics are.

1. *The Circle.* The circle is the only curve having the property that from a special point inside the curve (its center) a ball shot in any direction will return to that point.

This one is obvious. Did you know Lewis Carroll had a circular billiard table?

2. *The Ellipse.* Any ball shot from one focus of an ellipse will pass through the other focus after one bounce.

The converse is also true; any smooth curve having two special points with the stated reflection property must be an ellipse and those two special points its foci. Although the proof that a ball shot from one focus bounces through the other is easily given using derivatives, the following proof uses no calculus tools (See "A Pretrigonometry Proof of the Reflection Property of the Ellipse," by Zalman Usiskin in *The College Mathematics Journal* (November 1986), 418.

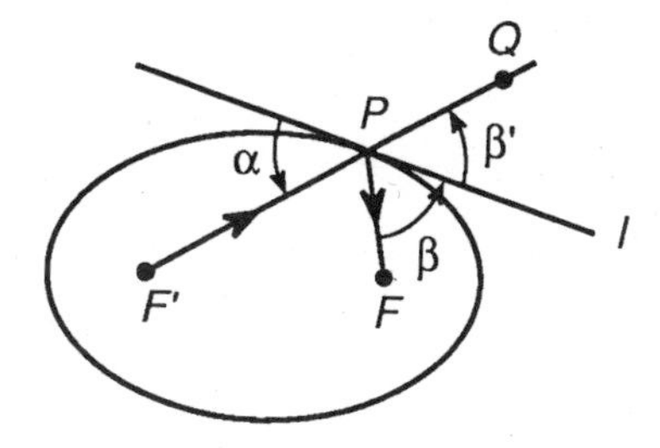

We must show that the two angles α and β made with tangent ℓ at P in our figure are equal.

Let Q be the reflection of F over line ℓ, so $FP = PQ$ and $\beta = \beta'$. Now of all points on ℓ, P is the unique point that minimizes the sum of the distances $F'P + FP$ (since other points P' on ℓ lie on other ellipses with foci F and F' corresponding to larger sums $F'P' + P'F$). Since $PF = PQ$, P is also chosen on ℓ so that the sum $F'P + PQ$ is minimal. This requires that F, P, and Q be collinear, so $\alpha = \beta'$; but $\beta' = \beta$ so $\alpha = \beta$ and the reflective property is established.

3. *The Parabola.* A ray entering a parabola parallel to the axis will be reflected through the focus.

The equation of the parabola with focus $(0, a)$ and directrix $y = -a$ is $y = \frac{1}{4a}x^2$. Let line ℓ be tangent to the curve at (x_0, y_0). We will show $\alpha = \beta$; that is, a ray parallel to the axis of the parabola is reflected through the focus.

The slope of ℓ is $\frac{x_0}{2a}$ (the derivative of $\frac{1}{4a}x^2$ at x_0). Therefore the equation of line ℓ is $y = y_0 + \frac{x_0}{2a}(x - x_0)$, which has y intercept $y_0 - \frac{x_0^2}{2a} = y_0 - 2y_0 = -y_0$. This means that quadrilateral $FPQS$ is a rhombus: $FS = y_0 + a = PQ$ and $FP = PQ$ by definition of the parabola. Since the diagonal $\overline{PS}$ of the rhombus bisects the corner angles, we see that $\beta = \alpha' = \alpha$, as required.

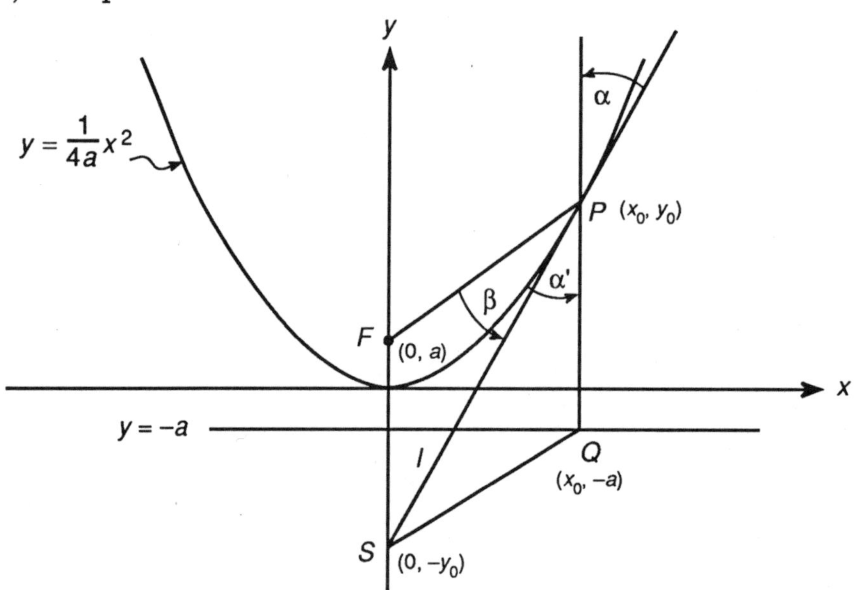

4. *The Hyperbola.* A ray passing through one focus of a hyperbola is reflected along a line passing through the other focus.

A proof can be given that is similar to the one given for the parabola.

A nice series of three articles on the conics is found in the UMAP Journal, a periodical devoted to undergraduate mathematics and its applications. The articles, all by Lee Whitt are: "The Standup Conic Presents: The Parabola and Applications," Vol III, No. 3, 1982; "The Standup Conic Presents: The Ellipse and Applications," Vol. IV, No. 2, 1983; "The Standup Conic Presents: The Hyperbola and Applications," Vol. V, No. 1, 1984.

Summary of the Reflection Properties of Conics

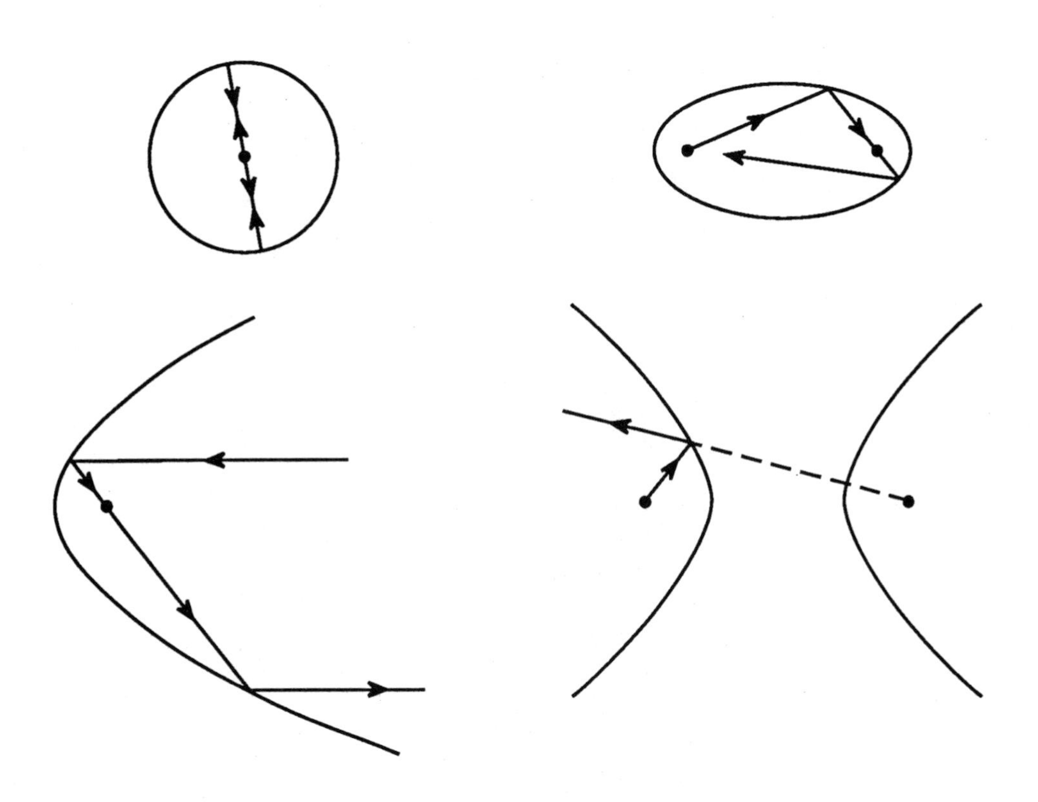

Exercises on Billiard Balls and Conic Sections

1. Explore the billiard properties of the curve obtained by placing a fixed-length loop of string around three non-collinear pegs, drawing it taut with a pencil, and drawing the resulting curve. If a ball passes over one of the points where a peg is, does every later segment of the ball's path contain one of the three points corresponding to a peg? Must the same be true of all previous segments?

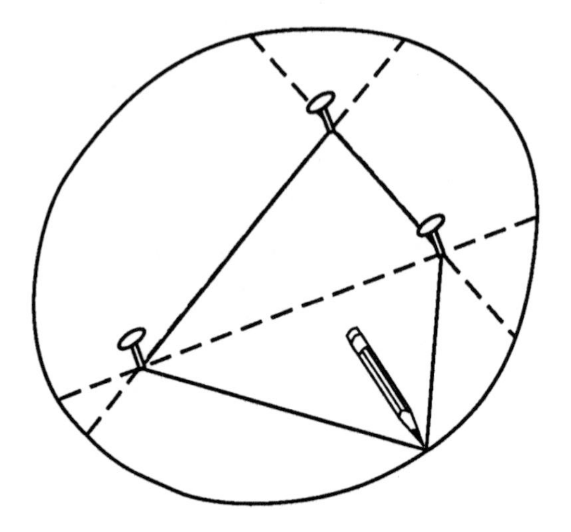

2. Examine the billiard properties of two parabolas that intersect as shown, with a common axis and common focus. In particular, what is the path of a ball shot from the focus?

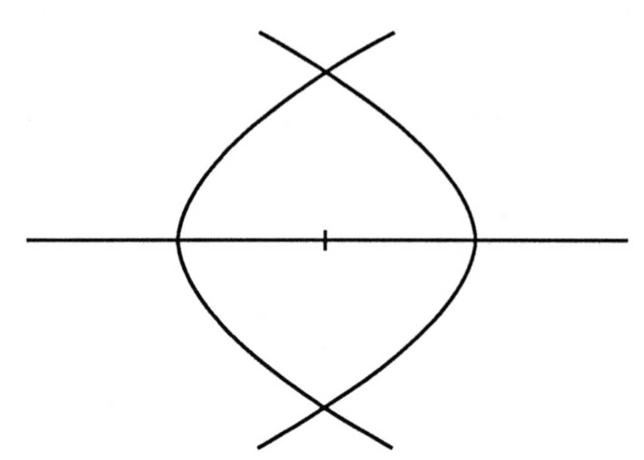

Answers to Exercises on Billiard Balls and Conic Sections

1. The curve is made up of six arcs of three different ellipses. Using the reflective properties of ellipses and limit arguments, one can see the curve is smooth; that is, it has a continuously turning tangent. (There are no "corners" where the different elliptical segments meet.) The answers to both questions are yes. (For previous segments, simply direct the ball the other way.)

2. Any ball shot from the focus (neither horizontally nor vertically) repeatedly traces out two congruent triangles, each of which has a horizontal side.

BIBLIOGRAPHY

In the following pages are some of the authors' favorite books on calculus, and a selection of articles that have appeared, since 1970, in one of the following journals: The American Mathematical Monthly, The College Mathematics Journal, Mathematics Gazette, Mathematics Magazine, The Mathematics Teacher, The Two Year College Mathematics Journal (renamed the College Mathematics Journal in 1980), and The UMAP Journal.

The articles are arranged topically, according to the table of contents of this handbook. Nearly any college library would have most of these books and journals in its collection, and so they are readily accessible. This bibliography should be a useful resource to the teacher for extra background reading and lesson planning, and should also be helpful to students for individual and group projects.

Books

Tom M. Apostol, *Calculus*, Volume 1, Blaisdell, 1967.

Tom M. Apostol, et al, ed., *Selected Papers on Calculus*, The Mathematical Association of America, 1969.

R. P. Boas, Jr., *A Primer of Real Functions*, Third edition, The American Mathematical Association, 1982.

Barry Cipra, *Misteaks...and how to find them before the teacher does*, 2nd ed., Academic Press, 1989.

Ivan Niven, *Maxima and Minima Without Calculus*, The American Mathematical Association, 1982.

Michael Spivak, *Calculus*, W. A. Benjamin, 1967.

Otto Toeplitz, *The Calculus: A Genetic Approach*, University of Chicago Press, 1963.

Journal Articles

The American Mathematical Monthly	=	Amer. Math. Monthly
The College Mathematics Journal	=	College Math. J.
Mathematics Gazette	=	Math. Gaz.
Mathematics Magazine	=	Math. Mag.
The Mathematics Teacher	=	Math. Teacher
The Two Year College Mathematics Journal	=	Two Year College Math. J.
The UMAP Journal	=	UMAP

GRAPHING

Thomas B. Baker, Sketching the Curve $y = x \sin \frac{1}{x}$, Math. Teacher 72 (Feb. 1979), 129-132.

A. A. Ball, Identifying Points of Inflection, Math. Gaz. 63 (Dec. 1979), 225-229.

R. J. Clark, A Limacon Property, Math. Gaz. 70 (Dec. 1986), 302-303.

Alice W. Essary, Rectangular Aids for Polar Graphs, Two Year College Math. J. 13 (June 1982), 200-203.

P. J. Giblin, What is an Asymptote? Math. Gaz. 56 (Dec. 1972), 274-284.

C. R. Haines, How Rough is a Rough Sketch? Math. Gaz. 63 (Dec. 1979), 265-267.

Keith Hirst, Composite Functions and Graph Sketching, Math. Gaz. 72, (June 1988), 114-117.

K. C. May, An Exercise in Polar Coordinates, Math. Gaz. 65 (Dec. 1981), 277-279.

Mark E. Saul, $\sin^2 x$: A Sheep in Wolf's Clothing, College Math. J. 21 (Jan. 1990), 43-44.

LIMITS

Gail H. Alneosen, The Schwarz Paradox: An Interesting Problem for the First-year Calculus Student, Math. Teacher 65 (Mar. 1972), 281-284.

John Baylis, $x^n e^{-x} \to 0$ as $x \to \infty$ Math. Gaz. 69 (Mar. 1985), 32-36.

William B. Gearhart and Harris S. Schultz, The Function $\frac{\sin x}{x}$, College Math. J. 21 (Mar. 1990), 90-99.

Sudhir K. Goel and Dennis M. Rodriquez, A Note on Evaluating Limits Using Riemann Sums, Math. Mag. 60 (Oct. 1987), 225-228.

Larry King, The G-S Connection, Two Year College Math. J. 14 (Jan. 1983), 42-47.

Tom Lehrer, There's a Delta for Every Epsilon, Amer. Math. Monthly 81 (June-July 1974), 612.

Sally Irene Lipsey and Wolfe Snow, The Appreciation of Radian Measure in Elementary Calculus, Math. Teacher 56 (Jan. 1973), 31-32.

P. Ramankutty and M. K. Vamanamurthy, Limit of the Composite of Two Functions, Amer. Math. Monthly 82 (Jan. 1975), 63-64.

Louis M. Rotando and Henry Korn, The Indeterminant For 0^0, Math. Mag. 50 (Jan. 1977), 41-42.

P. Shiu, Power Versus Exponential, Math. Gaz. 73 (Mar. 1989), 25-28.

John T. Varner III, Discovering a Calculus Theorem, Two Year College Math. J. 8 (Nov. 1977), 304.

INTERMEDIATE VALUE THEOREM

S. L. Tabachnikov, Considerations of Continuity, Quantum, (May 1990), 8-12.

THE DERIVATIVE

F. M. Arscott, Be Careful When You Dodge Around Newton, Math. Gaz. 63 (Oct. 1979), 169-173.

D. F. Bailey, Differentials and Elementary Calculus, College Math. J. 20 (Jan. 1989), 53.

R. P. Boas, Jr., L'Hôpital's Rule Without Mean-Value Theorems, Amer. Math. Monthly 76, (Nov. 1969), 1051-1053.

Joseph Browne and Jay D. Cook, An Intuitive Approach to L'Hôpital's Rule, Math. Teacher 75 (Dec. 1982), 757-758.

Robert J. Bumcrot, Some Subtleties in L'Hôpital's Rule, College Math. J. 15 (Jan. 1984), 51-52.

Barry A. Cipra, The Derivatives of the Sine and Cosine Functions, College Math. J. 18 (Mar. 1987), 139-141.

H. R. Corbishley, Improving Direct Iteration, Math. Gaz. 56 (May 1972), 110-113.

H. Martyn Cundy, $x^y = y^x$: An Investigation, Math. Gaz. 71 (June 1987), 131-135.

J. A. Eidswick, The Differentiability of a^x, Amer. Math. Monthly 82 (May 1975), 505-506.

Nathaniel A. Friedman, A Picture for the Derivative, Amer. Math. Monthly 84 (June-July 1977), 470-471.

Men-Chang Hu and Ju Kwei Wang, On the L'Hôpital Rule for Indeterminant Forms $\frac{\infty}{\infty}$, Math. Mag. 44 (Sept.-Oct. 1971), 217-218.

Xun-Cheng Huang, A Discrete, L'Hôpital's Rule, College Math. J. 19 (Sept. 1988).

J. P. King, L'Hôpital's Rule and the Continuity of the Derivative, Two Year College Math. J. 10 (June 1979), 197-198.

Roland E. Larson, Continuous Deformation of a Polynomial into its Derivative, Two Year College Math. J. 5 (Sept. 1974), 68-70.

Tom Lehrer, The Derivative Song, Amer. Math. Monthly 81 (May 1974), 490.

D. Mackie and T. Scott, Pitfalls in the Use of Computers for the Newton-Raphson Method, Math. Gaz. 69, (Dec. 1985), 252-257.

Raymond V. Morgan and Tony T. Warnock, Derivatives on the Hand-held Calculator, Math. Teacher 71 (Sept. 1978), 532-537.

Ivan Niven, Which is Larger e^π or π^e? Two Year College Math. J. 3 (Fall 1972), 13-15.

Michael Olinick and Bruce B. Peterson, Darboux's Theorem and Points of Inflection, Two Year College Math. J. 7 (Sept. 1976), 5-9.

W. C. Rheinboldt, Algorithms for Finding Zeros of Functions, UMAP 2, No. 1 (1981), 43-72.

David Rudd, A Closer Look at an Advanced Placement Calculus Problem, Math. Gaz. 78 (Apr. 1985), 288-291.

Harry Sedinger, Derivative Without Limits, Two Year College Math. J. 11 (Jan. 1980), 54-55.

David Tall, The Blancmange Function: Continuous Everywhere but Differentiable Nowhere, Math. Gaz. 66 (Mar. 1982), 11-22.

INVERSE FUNCTIONS

R. P. Boas, Jr., Inverse Functions, College Math J. 16 (Jan. 1985), 42-47.

R. P. Boas, Jr., and M. B. Marcus, Inverse Functions and Integration by Parts, Amer. Math. Monthly 81 (Aug.-Sept. 1974), 760-761.

Neal C. Raber and Richard J. Turek, Graphing Inverse Functions, Math. Teacher 72 (Apr. 1979), 276-277.

Ernst Snapper, Inverse Functions and Their Derivatives, Amer. Math. Monthly 97 (Feb. 1990), 144-147.

MAXIMUM-MINIMUM

Herbert Bailey, A Surprising Max-Min Result, College Math J. 18 (1987), 225-229.

Jean H. Bevis, County Agent's Problem: Or, How Long is a Short Barn? Math. Teacher 77 (Apr. 1984), 278-282.

James N. Boyd, Rectangles with Weighted Sides, Math. Teacher 74, (Jan. 1981), 36-38.

John W. Dawson, Jr., Hanging a Birdfeeder: Food for Thought, College Math. J. 21 (Mar. 1990), 129-130.

Mona Fabricant, A Classroom Discovery in High School Calculus, Math. Teacher 65 (Dec. 1972), 744-745.

Harley Flanders, Analysis of Calculus Problems, Math. Teacher 65 (Jan. 1972), 9-12.

T. Michael Flick, Encounter with Introductory Calculus, Math. Teacher 74 (Oct. 1981), 546-547.

Edwin C. Gibson and Jane B. Gibson, I Can See Clearly Now (Another look at Norman windows), Math. Teacher 75 (Nov. 1982), 694-696.

Dale T. Hoffman, Smart Soap Bubbles Can Do Calculus, Math. Teacher 72 (May 1979), 377-385, 389.

Pado Montuchi and Warren Page, Behold: Two Extremum Problems (and the Arithmetic-Geometric Mean Inequality), College Math. J. 19 (Sept. 1988), 347.

Norman Schaumberger, The AM-GM Inequality via $x^{\frac{1}{x}}$, College Math. J. 20 (Sept. 1989), 320.

Linda G. Thompson, A Second Look at the Second Derivative Test for Extreme Points, Math. Teacher 76 (May 1983), 336-337.

W. Thurmon Whitley, The Lifeguard Problem, UMAP 10, No. 2 (1989), 179-187.

MEAN VALUE THEOREM

R. P. Boas, Jr., Traveler's Surprises, Two Year College Math. J. 10 (Mar. 1979), 82-88.

__________, Who Needs those Mean-Value Theorems Anyway? Two Year College Math. J. 12 (June 81), 178-181.

Kenneth Goldberg, The Mean Value Theorem: Now You C it, Now You Don't, Math. Teacher 69 (Apr. 1976), 271-274.

P. Ivády, An Application of the Mean Value Theorem, Math Gaz. 67 (June 1983), 126-127.

Mary Powderly, A Geometric Proof of Gauchy's Generalized Law of the Mean, Two Year College Math. J. 11 (Nov. 1980), 329-330.

Thomas W. Shilgalis, Using Discovery in the Calculus Class, Math. Teacher 68 (Feb. 1975), 144-147.

Robert S. Smith, Rolle Over Lagrange–Another Shot at the Mean Value Theorem, College Math. J. 17 (Nov. 1986), 403-406.

RELATED RATES

Joe Dan Austin, How Fast Can You Watch? Math. Teacher 73 (Apr. 1980), 262-263.

RIEMANN SUMS

M. Bridger, Simpson and the Circle, Math. Gaz. 61 (Oct. 1977), 216-217.

Frank Burk, Behold! The Midpoint Rule is Better Than the Trapezoidal Rule for Concave Functions, College Math. J. 16 (Jan. 1985), 56.

___________, Archimedes' Quadrature and Simpson's Rule, College Math. J. 18 (May 1987), 222-223.

James D. Fabrey, The Power Formula for Integrals, Math. Gaz. 60 (June 1976), 134-135.

H. M. Finucan, An Elementary Proof of the Simpson Error, Math. Gaz. 60 (Mar. 1976), 63-64.

T. Kiang, An Old Chinese Way of Finding the Volume of a Sphere, Math. Gaz. 56 (Feb. 1972), 88-91.

John Mills, Numerical Integration: A Teaching Approach, Math. Gaz. 65 (Mar. 1981), 1-5.

A. Orton, Teaching $\int_1^a \frac{1}{x} dx = \ln a$ Math. Gaz. 72 (Dec. 1988), 271-276.

C. W. Puritz, Area and Volume of a Sphere, Math. Gaz. 57 (Oct. 1973), 206-207.

Arthur Richert, A Non-Simpsonian Use of Parabolas in Numerical Integration, Amer. Math. Monthly 92 (1985), 425-426.

Henry J. Schultz, The Sum of the kth Powers of the First n Integers, Amer. Math. Monthly 87 (June-July 1980), 478-481.

R. E. W. Shipp, A Simple Derivation of the Error in Simpson's Rule, Math. Gaz. 54 (Oct. 1970), 292-293.

S. K. Stein, The Error of the Trapezoidal Method for a Concave Curve, Amer. Math. Monthly 83 (Oct. 1976), 643-645.

ANTIDIFFERENTIATION

G. L. Alexanderson and L. F. Klosinski, Some Surprising Volumes of Revolution, Two Year College Math. J. 6 (Sept. 1975), 13-15.

Ashok K. Arora, Sudhir K. Goel, and Dennis M. Rodriguez, Special Integration Techniques for Trigonometric Integrals, Amer. Math. Monthly 95 (Feb. 1988), 126-130.

Frank Burk, $p/4$ and $\ln 2$ Recursively, College Math. J. 18 (Jan. 1987), 51.

F. Chorlton, Some Infinite Integrals, Math. Gaz. 64 (Oct. 1980), 190-193.

Robert C. Crawford, Using Inverse Functions in Integration, Two Year College Math. J. 8 (Mar. 1977), 107-109.

John W. Dawson Jr., Contrasting Examples in Improper Integration, Math. Teacher 83 (Mar. 1990), 201-202.

A. D. Fitt, What They Don't Teach You About Integration at School, Math. Gaz. 72 (Mar. 1988), 11-15.

John Frohliger and Rich Poss, Just an Average Integral, Math. Mag. 62 (Oct. 1989), 260-261.

Ann D. Holley, Integration by Geometric Insight – a Student's Approach, Two Year College Math. J. 12 (Sept. 1981), 268-270.

Arnold J. Insel, A Direct Proof of the Integral Formula for Arctangent, College Math. J. 20 (May 1989), 235-236.

Joscelyn A. Jarrett, Integrating the Inverse of a Function Whose Integral is Known, Math. Teacher 80 (Feb. 1987), 118-120.

Herbert E. Kasube, A Technique for Integration by Parts, Amer. Math. Monthly 90 (Mar. 1983), 210-211.

James McKim, The Problem of Galaxa: Infinite Area Versus Finite Volume, Math. Teacher 74 (Apr. 1981), 194-296.

V. N. Murty, How Close are the Riemann Sums to the Integral They Approximate? Two Year College Math. J. 11 (Sept. 1980), 268-270.

George P. Richardson, Reconsidering Area Approximations, Amer. Math. Monthly 95 (Oct. 1988), 754-757.

Edward Rozema, Estimating the Error in the Trapezoidal Rule, Amer. Math. Monthly 87 (Feb. 1980), 124-128.

R. Rozen and A. Sofo, Area of a Parabolic Region, College Math. J. 16 (Nov. 1985), 400-402.

Norman Schaumberger, The Evaluation of $\int_a^b \frac{dx}{x^2}$ and $\int_b^a \frac{dx}{\sqrt{x}}$, Math. Teacher 64 (Nov. 1971), 605-606.

A. H. Schoenfeld, The Curious Substitution z = tan q/2 and the Pythagorean Theorem, Amer. Math. Monthly 84 (May 1977), 370-372.

Lee L. Schroeder, Buffon's Needle Problem: An Exciting Application of Many Mathematical Concepts, Amer. Math. Monthly 56 (Feb. 1974), 183-186.

James M. Sconyers, Approximation of Area Under a Curve, Math. Teacher 77 (Feb. 1984), 92-93.

W. A. Stannard, Applying the Technique of Archimedes to the "Birdcage" Problem, Math. Teacher 72 (Jan. 1979), 58-60.

S. K. Stein, Formal Integration: Dangers and Suggestions, Two Year College Math. J. 5 (Sept. 1974), 1-7.

A. E. Stratton, The Curious Substitution z = tan q/2 and the Pythagorean Theorem, Amer. Math. Monthly 86 (Aug.-Sept. 1979), 584-585.

W. Vance Underhill, Finding Bounds for Definite Integrals, College Math. J. 15 (Nov. 1984), 426-429.

J. Van Yzeren, Moirre's and Fresnel's Integrals by Simple Integration, Amer. Math. Monthly 86 (Oct. 1979), 691-693.

John T. Varner II, Comparing a^b and b^a Using Elementary Calculus, Two Year College Math. J. 7 (Dec. 1976), 46.

Robert Weinstock, Elementary Evaluations of $\int_0^\infty e^{-x^2}dx$, $\int_0^\infty \cos x^2 dx$, and $\int_0^\infty \sin x^2 dx$, Monthly 97 (Jan. 1990), 39-42.

André L. Yandle, A Note on Integration by Parts, College Math. J. 16 (Sept. 1985), 282-283.

FUNDAMENTAL THEOREM

R. A. B. Bond, An Alternative Proof of the Fundamental Theorem of Calculus, Math. Gaz. 65 (Dec. 1981), 288-289.

Gilbert Strang, Sums and Differences vs. Integrals and Derivatives, College Math. J. 21 (Jan. 1990), 20-27.

Charles Swartz and Brian S. Thomson, More on the Fundamental Theorem of Calculus, Amer. Math. Monthly 95 (Aug.-Sept. 1988), 644-648.

APPLICATIONS

Jeanne L. Agnew and James R. Choike, Transitions, College Math. J. 18 (Mar. 1987), 124-133.

Gerald D. Brazier, Calculus and Capitalism–Adam Smith Revisits the Classroom, Math. Teacher 71 (Jan. 1978), 65-67.

Frank Budden, Cassette Tapes, Math. Gaz. 63 (June 1979), 113-116.

Thomas W. Casstevens, Population Dynamics of Governmental Bureaus, UMAP 5, No. 2 (1984), 177-199.

Colin Clark, Some Socially Relevant Applications of Elementary Calculus, Two Year College Math. J. 4 (Spring 1973), 1-15.

Simon Cohen, Ascent-Descent, UMAP 1, No. 2 (1980), 87-102.

William P. Cooke, Building the Cheapest Sandbox: An Allegory with Spinoff, Math. Mag. 60 (Apr. 1987), 101-104.

John C. Hegarty, A Depreciation Model for Calculus Classes, College Math. J. 18 (May 1987), 219-221.

Brindell Horelick and Sinah Koont, A Strange Result in Visual Perception, UMAP 1, No. 1 (1980), 31-47.

___________, Selection in Genetics, UMAP 2, No. 1 (1981), 29-42.

__________, Radioactive Chains: Parents and Children, UMAP 10, No. 3 (1989), 217-235.

James F. Hurley, An Application of Newton's Law of Cooling, Math. Teacher 56 (Feb. 1874), 141-142.

M. S. Klamkin, Dynamics: Throwing the Javelin, Putting the Shot, UMAP 6, No. 2 (1985), 3-18.

Neal Koblitz, Problems That Teach the Obvious but Difficult, Amer. Math. Monthly 95 (Mar 1988), 254-257.

Steven Kolmes and Kevin Mitchess, Information Theory and Biological Diversity, UMAP 11, No. 1 (1990), 25-62.

Peter A. Lindstrom and Richard G. Montgomery, Common Wisdom, Logarithmic Differentiation, and Compound Interest, UMAP 10, No. 4 (1989), 21-24.

Anthony Lo Bello, The Volumes and Centroids of Some Famous Domes, Math. Mag. 61 (June 1988), 164-170.

Anthony Lo Bello and Maurice D. Weir, Glotto Chronology: An Application of Calculus to Linguistics, UMAP 3, No. 1 (1982), 83-99.

Elmo Moore and Charles M. Biles, Fever, UMAP 4, No. 3 (1983), 253-258.

Yves Nievergelt, Graphic Differentiation Clarifies Health Care Pricing, UMAP 9, No. 1 (1988), 51-86.

Thomas O'Neil, A Mathematical Model of a Universal Joint, UMAP 3, No. 2 (1982), 199-219.

__________, Constructing Power Lines, UMAP 4, No. 3 (1983), 259-264.

W. L. Perry, A Bifurcation Problem in First Semester Calculus, Two Year College Math. J. 14 (Jan. 1983), 57-60.

G. J. Porter, New Angles on an Old Game, Amer. Math. Monthly 88 (April 1981), 285-286.

John E. Prussing, The Relationship Between Directional Heading of an Automobile and Steering Wheel Deflection, UMAP 7, No. 1 (1986), 29-44.

N. Reed, Elementary Proof of the Area Under a Cycloid, Math. Gaz. 70 (Dec. 1986), 290-291.

Joseph V. Roberti, Interdimensional Relationships, Math. Teacher 81 (Feb. 1988), 96-100.

Jerome Rosenthal, An Introduction to Hyperbolic Functions in Elementary Calculus, Math. Teacher 79 (Apr. 1986), 298-300.

Arthur. C. Segal, Some Transcendental Results Concerning Air Resistance, UMAP 7, No. 1 (1986), 19-28.

Alan H. Schoenfeld, The Curious Fate of an Applied Problem, College Math. J. 20 (Mar. 1989), 115-123.

Keith A. Struss, Exploring the Volume-Surface Area Relationship, College Math. J. 21 (Jan. 1990), 40-43.

Rudy Svoboda, The Calculus of Leaves: A Modelling Project for a Calculus Class, UMAP 4, No. 1 (1983), 19-27.

William V. Thayer, The St. Louis Arch Problem, UMAP 4, No. 3 (1983), 347-365.

Bert K. Waits and Jerry L. Silver, A New Look at an Old Work Problem, Two Year College Math. J. 4 (Fall 1973), 53-55.

SEQUENCES

D. H. Armitage, Two Applications of Bernoulli's Inequality, Math. Gaz. 66 (Dec. 1982), 309-310.

John C. Egsgard, An Interesting Introduction to Sequences and Series, Math. Teacher 81 (Feb. 1988), 108-111.

O. Iden, A Sum from an Integral, Math. Gaz. 68 (Mar. 1984), 51-52.

R. F. Johnsonbaugh, Another Proof of an Estimate for e, Amer. Math. Monthly 81 (Nov. 1974), 1011-1012.

John Pym, An Approach to Convergence, Math. Gaz. 60 (Mar. 1976), 57-59.

SERIES

A. L. Andrew, The Mathematics of Population Growth, Math. Gaz. 58 (Dec. 1974), 272-276.

D. H. Armitage, A Derivation of the Real Exponential Series, Math. Gaz. 63 (June 1979), 123-124.

M. D. Atkinson, How to Compute the Series Expansions of sec x and tan x, Amer. Math. Monthly 93 (May 1986), 387-389.

Richard Beigel, Rearranging Terms in Alternating Series, Math. Mag. 54 (Nov. 1981), 244-246.

R. P. Boas, Jr., Partial Sums of Infinite Series, and How They Grow, Amer. Math. Monthly 84 (Apr. 1977), 237-258.

Fred Brauer, A Simplification of Taylor's Theorem, Amer. Math. Monthly 94 (May 1987), 453-454.

F. Chorlton, Outpourings from a Trigonometric Function, Math. Gaz. 71 (Dec. 1987), 305-307.

Sydney C. K. Chu and Man Keung Siu, How Far Can You Stick Out Your Neck? College Math. J. 17 (Mar. 1986), 122-132.

Teresa Cohen and William J. Knight, Convergence and Divergence of $\sum \frac{1}{n^p}$, Math. Mag. 52 (May 1979), 178.

Curtis Cooper, Geometric Series and a Probability Problem, Amer. Math. Monthly 93 (Feb. 86), 126-127.

C. C. Cowen, K. R. Davidson, and R. P. Kaufman, Rearranging the Alternating Harmonic Series, Amer. Math. Monthly 87 (Dec. 1980), 817-819.

H. Martyn Cundy, et al, $4\left(1 - \frac{1}{3} + \frac{1}{5} - \frac{1}{7} + \cdots\right)$, Math. Gaz. 67 (Oct. 1983), 171-188.

K. A. Deadman, Convergence of Geometric Series, Math. Gaz. 54 (May 1970), 140-141.

Miltiades S. Demos, Class Notes on Series Related to the Harmonic Series, Math. Mag. 46 (Jan. 1973), 40-41.

J. E. Drummond, On Stacking Bricks to Achieve a Large Overhang, Math. Gaz. 65 (Mar. 1981), 40-42.

Mark Finkelstein, $\ln 2 = 1 - \frac{1}{2} + \frac{1}{3} - \frac{1}{4} + \cdots$, Amer. Math. Monthly 94 (June-July 1987), 541-542.

Joscelyn A. Jarrett, Regular Polygons and Geometric Series, Math. Teacher 75 (Mar. 1982), 258-261.

Wells Johnson, Power Series Without Taylor's Theorem, Amer. Math. Monthly 91 (June-July 1984), 367-369.

Richard Johnsonbaugh, Applications of Calculators and Computers to Limits, Math. Teacher 69 (Jan. 1976), 60-65.

Gerald Jungck, An Alternative to the Integral Test, Math. Mag. 56 (Sept. 1983), 232-234.

Franklin Kemp, Infinite Series Flow Chart for $\sum a_n$, Two Year College Math. J. 13 (June 1982), 199.

Benjamin G. Klein, Proof Without Words: $1 + r + r^2 + \cdots = \frac{1}{1-r}$, Math. Mag. 61 (Oct. 1988), 219.

Frank Kost, A Geometric Proof of the Formula for ln 2, Math. Mag. 44 (Jan. 1971), 37-38.

I. E. Leonard and James Duemmel, More - and Moore - Power Series Without Taylor's Theorem, Amer. Math. Monthly 92 (Oct. 1985), 588-589.

Robert Li, An Easier Series Counter-Example, Math. Gaz. 70 (Oct. 1986), 220.

A. R. G. Macdivitt and Yukio Yanagisawa, An Elementary Proof That e is Irrational, Math. Gaz. 71 (Oct. 1987), 217.

Charles C. Mumma, II, N! and the Root Test, Amer. Math. Monthly 93 (Aug-Sept. 1986), 561.

Roger B. Nelsen, Proof Without Words: Differentiated Geometric Series, Math. Mag. 62 (Dec. 1989), 332-333.

Esteban I. Poffald, The Remainder in Taylor's Formula, Amer. Math. Monthly 97 (Mar. 1990), 205-213.

P. J. Rippon, Convergence with Pictures, Amer. Math Monthly 93 (June-July 1986), 476-478.

J. Rubinstein, Infinite Geometric Series – a child's view, Math. Teacher 74 (Oct. 1981), 534-535.

Richard H. Schwartz, Population Growth, Tree Diagrams, and Infinite Series, UMAP 6, No. 1 (1985), 35-40.

Peter Shiu, A Generalization of Some Convergence Tests, Math. Gaz. 56 (Oct. 1972), 227-228.
__________, How Slowly Can a Series Converge?, Math. Gaz. 56 (Dec. 1972), 285-288.

Philip Tuchinsky, Zipf's Law and His Efforts to Use Infinite Series in Linguistics, UMAP 1, No. 1 (1980), 80-100.

R. Vyborny, Differentiation of Power Series, Amer. Math. Monthly 94 (April 1987), 369-70.

A. D. Wadwha, An Interesting Subseries of the Harmonic Series, Amer. Math. Monthly 82 (Nov. 1975), 931-933.

__________, Some Convergent Subseries of the Harmonic Series, Amer. Math. Monthly 85 (Oct. 1978), 661-663.

A. J. B. Ward, Divergence of the Harmonic Series, Math. Gaz. 54 (Oct. 1970), 277.

Michael Worboys, The Geometric Series – a Geometric Demonstration, Math. Gaz. 60 (Oct. 1976), 204-205.

Robert M. Young, The Error in Alternating Series, Math. Gaz. 69 (June 1985), 120-121.

CONIC SECTIONS

Hermann Von Baravalle, Conic Sections in Relation to Physics and Astronomy, Math. Teacher 63 (Feb. 1970), 101-109

James H. Foster and Jean J. Pedersen, On the Reflective Property of Ellipses, Amer. Math. Monthly 87 (Apr. 1980), 294-297.

R. W. Payne, The Reflector Property of the Ellipse, Math. Gaz. 58 (Oct. 1974), 215.

Lee Whitt, The Standup Conic Presents: The Parabola and Applications, UMAP 3, No. 3 (1982), 285-313.

__________, The Standup Conic Presents: The Ellipse and Applications, UMAP 4, No. 2 (1983), 157-183.

__________, The Standup Conic Presents: The Hyperbola and Applications, UMAP 5, No. 1 (1984), 9-21.

Robert C. Williams, A Proof of the Reflective Property of the Parabola, Amer. Math. Monthly 94, (Aug.-Sept. 1987), 667-668.

GENERAL

R. P. Boas, Jr., "A Primer of Real Functions," *Third Edition , Carus Mathematical Monograph Nol. 13*, Mathematical Association of America (1982)

E. F. Assmus, Jr., Pi, Amer. Math. Monthly 92 (Mar. 1985), 213-214.

R. P. Boas, Jr., Calculus as an Experimental Science, Two Year College Math. J. 2 (Spring 1971), 36-39.

Carl B. Boyer, The History of the Calculus–an Overview, Two Year College Math. J. 1 (Spring 1970), 60-86.

Frank Burk, Mean Inequalities, Two Year College Math. J. 14 (Nov. 1983), 431-434.

Thomas Butts, Fixed Point Iteration–An Interesting Way to Begin a Calculus Course, Two Year College Math. J. (Jan. 1981), 2-7.

Robert Decker, Discovering Calculus, Math. Teacher 82 (Oct. 1990), 558-563.

Henry C. Finlayson, The Place of ln x Among the Powers of x, Amer. Math. Monthly 94 (May 1987), 450.

Judith V. Grabiner, The Changing Concept of Change: The Derivative From Fermat to Weierstrass, Math. Mag. 56 (Sept. 1983), 195-206.

H. N. Gupta, A Simple Proof That $e < 3$, Math. Gaz. 62 (June 1965), 124.

Jaroslav Hanel, A Simple Proof of the Irrationality of π^4, Amer. Math. Monthly 93 (May 1986), 374-375.

Jim Howard and Joe Howard, Equivalent Inequalities, College Math. J. 19 (Sept. 1988), 350-352.

Morris Kline, Euler and Infinite Series, Math. Mag. 56 (Nov. 1983), 307-315.

Hans Liebeck, Introducing Hyperbolic Functions, Math. Gaz. 71 (March 1987), 55-56.

E. R. Love, Some Logarithm Inequalities, Math. Gaz. 64 (Mar. 1980), 55-57.

Coreen L. Mett, Writing as a Learning Device in Calculus, Math. Teacher 80 (Oct. 1987), 534-537.

George McCarty, Calculator-Demonstrated Math Instruction, Two Year College Math. J. 11 (Jan. 1990), 42-47.

Allen E. Parks, π, e, and Other Irrational Numbers, Amer. Math. Monthly 93 (Nov. 1986), 722-723.

Jean Pedersen and Peter Ross, Testing Understanding and Understanding Testing, College Math. J. 16 (June 1985), 178-185.

Donald E. Richmon, An Elementary Proof of a Theorem in Calculus, Amer. Math. Monthly 92 (Oct. 1985), 589-590.

V. Frederick Rickey, Isaac Newton: Man, Myth, and Mathematics, College Math. J. 18 (Nov. 1987), 362-389.

H. Samelson, To e via Convexity, Amer. Math. Monthly 81 (Nov. 1974), 1012-1013.